APPLIED MANAGING FOR ENTREPRENEURSHIP

APPLIED MANAGING FOR ENTREPRENEURSHIP

Walter Amedzro St-Hilaire, PhD

Apple Academic Press Inc.
4164 Lakeshore Road
Burlington ON L7L 1A4
Canada

Apple Academic Press, Inc.
1265 Goldenrod Circle NE
Palm Bay, Florida 32905
USA

© 2021 by Apple Academic Press, Inc.

First issued in paperback 2021

Exclusive co-publishing with CRC Press, a Taylor & Francis Group
No claim to original U.S. Government works

ISBN 13: 978-1-77463-501-8 (pbk)
ISBN 13: 978-1-77188-911-7 (hbk)

Library and Archives Canada Cataloguing in Publication
Title: Applied managing for entrepreneurship / Walter Amedzro St-Hilaire, PhD.
Names: Amedzro St-Hilaire, Walter, author.
Description: Includes bibliographical references and index.
Identifiers: Canadiana (print) 20200267272 | Canadiana (ebook) 20200267701 | ISBN 9781771889117 (hardcover) | ISBN 9781003034377 (ebook)
Subjects: LCSH: Industrial management. | LCSH: Organizational effectiveness.
Classification: LCC HD31.2 .A44 2020 | DDC 658—dc23

CIP data on file with US Library of Congress

Apple Academic Press also publishes its books in a variety of electronic formats. Some content that appears in print may not be available in electronic format. For information about Apple Academic Press products, visit our website at **www.appleacademicpress.com** and the CRC Press/Routledge website at **www.routledge.com**

About the Author

Walter Amedzro St-Hilaire, PhD

Prof. Dr. Walter Amedzro St-Hilaire is a research leader. He is the author of more than 15 books and around 30 scientific articles. His specialization areas include portfolio management (bank, telecom, health, energy, and agribusiness), project management, entrepreneurship policies, corporate governance, business technology, strategic management, business economics, risk management, economic infrastructures, public administration, international development, and applied economics. Prof. Amedzro St-Hilaire has to his credit several years of academic experience, having taught at various universities: HEC-Montreal (Canada), University of Ottawa (Canada), Northwestern University (USA), and George Washington University (USA). He is also Projects Economics and Financial Business Expert for several institutions and international organizations. Proud holder of several postdoctoral awards, he directs the *Global Journal of Strategies & Governance* and the *Management & Applied Economics Review*. The richness of his academic and professional experience gives him a certain expertise in the research of the highest spheres of decision-making as well as in education.

Contents

Abbreviations

ERP	enterprise resource planning
PIMS	profit impact of market strategy
R&D	research and development
ROI	return on investment
SMEs	small and medium-sized enterprises
SWOT	strengths, weaknesses, opportunities, threats

Acknowledgment

The author thanks FordBridge University, ExpertActions ExiGlobal Group, Northwestern University and the Chair of Institutional Governance & Strategic Leadership Research for funding this research.

Foreword

This book provides an overview of the approach and mechanisms with regard to optimizing management. It is remarkable indeed that the amount of research related to the optimization of organizations has increased significantly during recent years. Nevertheless, like all scientific disciplines, applied management is called to question and renew itself.

First of all, it should be pointed out that at the level of the advanced research and basic education, there is a very sharp criticism of the grids and models of applied analysis. Such was the case of the LCAG model and operational matrices, and finally the models of Porter. The questioning is about the deterministic nature of these mechanisms. Probably beyond the intention of their authors, often victims of their educational success (pedagogy necessarily simplifying the presentation of models and theories) and consultant. Most often, these analyses emphasize the two relationships with two (market share and profits, for example). In this respect, it was better to show that relations could be more complex and contingent.

In terms of "business optimization," the central question is the relationship, consistency (fit) between strategic positioning and competitive advantage, based on the company's capabilities. However, the authors tend to rely increasingly on the skills required for each company rather than on the characteristics of the business context. They advocate increasingly individualized case studies rather than statistical studies of large amounts of data. But unanimity is far from being achieved on methods; a large scientific legitimacy is still given to statistical studies. This questioning of the general model is exemplified in the particular fate of optimization of small businesses, the role of the entrepreneur, and the crucial problems of creation. But the models and analyses are likely to be caused to differentiate further.

At the "corporate optimization," other dimensions are more integrated; this is the case of the moral dimension, ethical, ecological. Also, industrial policy and market globalization are decisive. Porter stressed that the competitiveness of local businesses was largely due to differences in local skills, as well as the aid and regulatory systems. Porter insists that his scheme is a dynamic system. More generally, he joined the specialists of evolutionary theory to emphasize the need for a dynamic approach to decision optimization (e.g., the analysis in terms of cards remains largely static). One of the major projects of decision analysis in applied management matters is the economic globalization.

Globally, more than ever, applied management appears as an open discipline, changing. Hopefully that ensures new configurations, new paradigms that make simple problems that have appeared much more complex in the eyes of the neophyte.

Prof. Dr. Walter Amedzro St-Hilaire

Introduction

Applied management has made significant progress. Treated for a long time as business economics, management, or marketing, the discipline has become essential for the adaptation of organizations. This recognition was reflected by an increase in scientific and educational work. These researches resume templates, commonly taught in the United States theories and grids. The goal here is to write a book for voluntarily analytic and synthetic sizes. The idea is to make a synthesis of different currents developed in applied management, based on their classification as it has been prepared by specialists.

This classification is highly questionable, of course, but it will bring up an important point: largely based on instrumental rationality initially and on policy issues, decision optimization gradually interested in the problems of choice of activities, distinguishing the "corporate" of "business optimization." Subsequently, some researchers have increasingly criticized the patterns and grids made; abandoning procedures, quite ineffective in troubled times, the authors have tended to focus on collective decision-making in organizations, and individuals, from the entrepreneur strategist. Currently, the expressed wish would lead to approaches that incorporate the procedures and processes using grids.

This is why we propose from the description of the problem to present the multiplicity of designs and areas of analysis applied to address the notion of competitive environment, the organization issues. We are, thus, led to problems related to the operational decision, then the decision maker, and entrepreneurship to awareness applied analysis (cases board) through an integrative approach that we advocate. It is remarkable indeed to note that this course is quite widely in succession in time, over the last 30 years, theoretical perspectives. Currently, specialists are focused on one aspect or another, often depending on their original discipline (industrial economics, marketing, management, management control, etc.). Our ambition probably was to present various facets, very synthetically. We limited the maximum references to authors, to retain only the logical progression from one stage to another. We have often sought to adapt the most common models for homogenization.

This book is intended as a guide for those students, executives, or managers who wish to think of the dynamics of applied management and different mechanisms to optimize and implement in an organization. It was necessary to present the concept and the different aspects it covers. Although our design does not differ materially from what is commonly accepted, there are nuances and on which we insist that betray our own biases. Importantly, we live in a complex world, and the world of organizations is elusive. We then need instruments, heuristics, that help to decipher what is going on, but acceptable temporarily, until we have a better understanding of the phenomena. In its own way, the applied management is a robust tool because it does not focus on specific relationships. It is interested in the processes that lead to a better decision-making process. Research contributes, moreover, dramatically, to identify elements to consider when looking at these processes. But more importantly, optimization mechanisms are a powerful policy tool for practitioners because they help bring order in a chaotic world and act appropriately.

A synthesis of the literature in five forms were selected: optimization as managing organization–environment relationship; optimization as an extension officers; optimization as an expression of a community of persons; optimization as a hinge part; and optimization as building a competitive advantage. Here, the author focuses on the factors to be considered when making the analysis of the strategic position of an organization. It then provides in the book, the general framework of analysis that is often guided by an overall purpose, which can be a mission statement or a general and lasting statement that defines the organization and its purpose. It begins with an analysis of the dynamics of the business context and continues with the analysis of organizational capabilities that can be the source of its competitive advantages. This external and internal analysis allows to define objectives, which fit into the broader purpose of the organization. These are the objectives that we call "intelligence optimization."

The author goes further in analyzing the context of business. It describes the different types of environments and techniques to analyze them. It attaches particular importance to the analysis of competition. Traditional models, including the industrial economy popularized by Porter, are described and illustrated. It also discusses how we can analyze the sociopolitical context. Before beginning the analysis of capabilities and resources of the organization, it aims to understand what gives or may give the organization an advantage over its competitors and what is or may be a handicap in the rivalry between the organizations to its competitors. Different models, such as those in the value chain, and ideas from core competencies are discussed.

The book recounts the major mechanisms usually chosen by business and the important issues they raise at quarterback. The generic optimizations are addressed, but also the most commonly used maneuvers. Here it must be said that the applied management material formulation is based not only on the analysis of the business context and organizational resources. The choice suggests the context of that case and the capabilities of the organization should also consider the values and preferences of leaders and what is acceptable to society. This is why various aspects of the complexity of organizations are taken into account. We had to start by describing what a complex situation is and show that in a situation of complexity, cause-and-effect relationship is unclear and the constructive power of leaders to lead people in the desired direction is often reduced.

The book focuses on the applied management of the diversified businesses. It addresses the historical conditions that led to the great waves of mergers and acquisitions, the reasons that encourage companies to diversify their activities, the various existing mechanisms of diversification, and challenges related to the applied management of the diversified businesses. On the issue of global companies, the author discusses the globalization of markets, companies and industries and different analysis models to better understand the dynamics of globalization. Subsequently, he looks at challenges of applied management in these structures and operational maneuvers that can be used in a context of geographic dispersion.

Finally, the book presents analytical tools to help leaders achieve change in their organization. We must therefore remember that, intended primarily for students wishing to learn the applied management organizations avoiding scholarship, this book is also useful for practitioners and consulting companies eager to update their booming knowledge.

CHAPTER 1

The Concept of Entrepreneurship

The concept of the entrepreneur is one of the most controversial and meaningful in operational and strategic analysis. It is given different meanings, which account for most of the differences in evaluation, especially on performance. Moreover, it should provide a clear explanation of what is meant by "spirit of enterprise" or "entrepreneurial." These definitions should clarify the debates on the fundamental operational act, which is that of creating and starting new businesses. The contractor is the one who makes operational decisions. But the word was charged with meaning to develop the market economy in industrial capitalism and economic thought.

Initially, in the merchant capitalism, the entrepreneur takes part in business transactions (e.g., between the weavers and cloth merchants). The contractor industry supports product manufacturing, committing its capital to buy machinery and organize production. According to Adam Smith, the entrepreneur has above all the virtue of savings and capital mobilization. Jean-Baptiste Say attributed to the contractor the key role of organizing production, a "combination of factors," which justifies its profit (which also pays its capital). Thus, the entrepreneur can see that his role is underestimated by liberal economists: for them, the entrepreneur is content to bow to market in its pursuit of profit maximization.

The entrepreneur began to be rehabilitated by the American economists at the beginning of the century, in what is called the institutionalized approach. Veblen stresses the vital role of the entrepreneur, and he fears that he is overtaken by "engineers," that is to say, pure employees' managers. Knight justifies profit as a reward for the risk and uncertainty that supports the entrepreneur. Joseph Schumpeter assigns a major role to entrepreneurship in the development of capitalism. This spirit is reflected in the constant search for innovation of all kinds (product, technology, markets, organization, etc.), this constant wave of innovation explaining the dynamics of capitalism. However, for Schumpeter, entrepreneurship does not necessarily identify an individual, as is too often said; it may very well exist in large companies or nonprofit institutions. It is realized through innovation.

It must be said that contemporary strategic management mobilizes many operational mechanisms that place particular importance on planning, prioritization and formalization.The contractor is supplanted by the employee manager, as pointed by Barnard and Burnham. Small business and the "little boss" appear as outdated categories, because of the race to economies of scale and size. Large companies are supposedly more efficient, more innovative, and more profitable. The small company that manages to survive becomes a larger player and turns into an organization "management." The rehabilitation of the entrepreneur and the entrepreneurial spirit, which is embodied in the entrepreneur, will increase, encouraging a climate of particular studies related to optimization of management practice.

These include: the failure of the major organizations in their response to different crises; the emergence of new activities related to the third industrial revolution, in which small businesses and the "new entrepreneurs" play a decisive role; the new conditions of the international division of labor to develop new industries, or implanting old industries, wherein countries and regions have developed a policy of support for the creation of companies; the loss of legitimacy of big business; the net job creation coming from sectors dominated by small businesses, and moreover, they often seem to have better resisted the successive crises; and finally, entrepreneurship has appeared, with some naivety (or bad faith), as a means of finding work for the unemployed.

One could speak of an "entrepreneurial phenomenon," maintained by the media and charisma of some business leaders. The craze seems to have fallen. Lucidity has been replaced. The success of the entrepreneurial act now appears to be linked to controlling operational variables: contractors' capabilities, skills of its organization, depending on the activity and resources, the quality of the project, of its decisions, expressed in the business plan. This raises the question of the classification of the contractor. There are some commonly taught classifications. We must bear in mind that they include to varying degrees several dimensions. In the capitalist system, the contractor performs more or less these three actions:

(1) It generates capital in order to make them grow. Such is the case of the owner-manager. But many entrepreneurs aid in making money for others (family, venture capital firms, debt capital, etc.).

(2) It organizes production and adds value through functions that increase market value. But many entrepreneurs have only limited capacity of "managing" (ability to drive individuals, coordinates, predicts, controls).

(3) It innovates the market. But many business leaders work on some already innovative markets (where they can also earn a very good living).

In other words, we must beware of the mythical image of the entrepreneur "dynamic," "risk taker," strategist and adventurer at once. We tried, therefore, what are the social and psychological traits that distinguished entrepreneurs from other workers.

(a) The most commonly advanced psychological trait is the need of/ for achievement (N of A: need for personal fulfillment), according to the theory developed by Mac Clelland. The contractor would do both: he would prove and realize something, as an athlete who crosses the Atlantic rowing. We generally add the two needs: the need of/for power—that is to say, the need to exercise power over others (the contractor will mount an organization)—and the need of/for affiliation—that is to say, the need to be recognized and integrated into a medium (being entrepreneurs are seen as a social promotion). Each entrepreneur will perceive these needs differently (a craftsman will feel probably more N affiliate of the N of A). Many authors have sought more or less frequent psychological skills for entrepreneurs. These include, in bulk: taking calculated risks, self-confidence, acceptance of responsibility, long-term vision, acceptance of multiple solutions situations, tenacity, acceptance of failure and questioning, etc. Needless to say, those few people possess all these qualities together!

(b) Other authors were interested in the social background of the entrepreneur. The results were highly controversial. We can finally assume that:

(1) The entrepreneur "heir," who perpetuates a business or an existing business, will often be the eldest in the family, maybe from a social group firmly established in this type of activity (existence of a network) which is relatively stable.

(2) The entrepreneur "innovative," who develops business in new activities, will often be in conflict with the family culture, will instead be the youngest, with a high A of N, agreeing to be in a business context that is turbulent, and even hostile.

(3) Also, there are many types that do not make any reference to the conception of the entrepreneur. The earliest classification is Norman Smith, who, observing entrepreneurs, classified them as:

- Artisan: With relatively few technical and management skills, decided rather abruptly to create his business, for the sake of independence, or to find a job. He plays on a network of relationships and takes advantage of an opportunity that presents itself. We can say that this is an incremental and reactive decision.

- Optimist: The term is unsuitable, as Smith refers rather to creators who have matured their project, acquired the technical skills, management, capital and resources before launching their business. We can talk about proactive and deliberate decision.

This classification, although still cited, is highly questionable. There are many other types of creators; regarding the heads of existing companies, it appears that almost all, in certain sectors, would type "artisan," which removes any interest in the classification! Smith himself has virtually disowned. Drawing on studies conducted by Miles and Snow, which distinguish entrepreneurs "adapters" and "innovative" entrepreneurs, led to a classification that takes into account both the type of organization and the degree of innovation.

The larger the organization is structured, the more we can assume that the contractor will have management skills (what is sometimes called "professional contractor") often acquired as a result of management experience, together with a complementary training. Such a classification allows to better understand the entrepreneurial archetypes. We suggest another type of classification, based on the aspirations of the leaders.

These are three in number: the search for the sustainability of the case, looking for the independence of the capital or of the autonomy of decision (which is not the same thing), and the search for growth, whether proactive or reactive. From these three fundamental aspirations, we can identify two extreme types of entrepreneurs (in the sense of business leaders):

(1) The PIC is driven by the logic of essentially patrimonial share (the company must contribute to increasing the value of assets held by the individual and/or family). As a result, priority is given to the sustainability of the case (the transmission is a major problem) as well as financial independence (refusal to external partners or bank indebtedness). Growth is only accepted if it does not affect these priority aspirations. PLCs are driven by the logic of action based more on valuation than on capital accumulation. The main objective is growth, to the extent that it identifies with profitability.

(2) In contrast to the above is the second category: the CAP. The CAP rather works on other people's money (which increases by effect of debt leverage, its return on equity), but he wants to maintain the autonomy of decision (risk capital is for it an ideal formula). Finally, they do not particularly wish to perpetuate in an activity, because the entry of new competitors and product maturity lead to lower rates of return.

The PIC will improve in stable, mature business, while the CAP in young and rowdy activities; they differ in management style (well, the CAP will be more open about its business context and will do more marketing). There are many other types of the entrepreneur. However, they raise much criticism.

They are only extreme cases of representation, facing an extraordinary diversity of contractors. An entrepreneur learns as he/she ages, reacts to events and may therefore change his/her profile. Simple typologies tend to be false, while too complex typologies become unusable. In fact, researchers should propose a typology by type of operational problem studied.

However, these types have some educational value and are a valuable diagnostic tool in the first analysis. In particular, they have the merit to emphasize the fundamental role of the leader, especially in companies with personal or family management in the operational choices.

And what is the role of entrepreneurship in all this? For contractor entrepreneurs, as economic analysis has especially emphasized, entrepreneurship is a corporate function. It is diffused in all organizations; it is characterized by the ability to innovate, your calculated risk taking, the ability to conceive, organize, and carry out a project: this is the meaning given by authors such as Peter Drucker and Mark Casson. Often, an individual or a group will have more or less this entrepreneurship: some will be more visionary ("prospectors"), others rather organizers ("adapters").

But it is in this corporate act of creation that entrepreneurship becomes a realization. Now, the business creation process has continued to grow. The established companies were becoming smaller, with less and less capital to start, either by necessity (low resource creators) or opportunity (created mainly in the service sector, less demanding in capital structure). We can also estimate that the actual creation of small or very small businesses has been underestimated by official statistics, due to the development of a very large underground economy (undeclared businesses).

To solve the problem of unemployment, to revitalize the regions, public institutions have, in almost all countries of the world organized assistance to business creation systems—the main problem being to find "entrepreneurs," that is to say, people with: required skill set, even for minimal tasks (e.g., maintain records, make an estimate); psychological characteristics of a "temperamental profile" corresponding to qualifications; a viable project, that is to say based on a coherent analysis of the proposed activity (technological skills, existence of a potential market, product reliability), and a sound assessment of the resources needed.

As a result, a multitude of aids were introduced (we lived in France where several hundreds of aids were launched), and organizations were created at national and local level, hoping to generate and/or support the creation companies. The results were highly controversial. The essential discussions focus on the following.

It is not sure whether there is consistency between all these organizations. In particular, it is possible to establish a "wild" competition between cities or regions (or countries) seeking to attract at all costs for entrepreneurs.

The aid is too complex. Paradoxically, it has especially benefited the subsidiaries of large groups, or large SMEs (small and medium-sized enterprises), which could acquire the expertise to complete the paperwork.

The aid is ineffective. They are most often designed to help the creative well beyond the most difficult phase is the startup. They can do that arriving after the victory (or defeat).

The aids are expensive. Although they apparently reduce mortality (between two-thirds and half of companies die within the first five years), have they not been granted, precisely because the project was viable at the outset. However, the cost per job created in the "techno polis," "nurseries," and other "business parks" may seem excessive for companies handpicked and would have probably robbed of their own. On these points, it is up to economists to try to measure the actual effectiveness of these incentives for entrepreneurship. However, applied management specialists are asking other questions:

(1) First, can we predict the success of a business project? Hence, it is better to first understand the circumstances of the creation of a company. Shapero hypothesized a "creation by moving." Shapero highlights factors such as psychological (need of achievement, independence), sociological (breaks: dismissal, personal or professional disappointment, etc.), and social (environment, home environment) as initial, potential conditions. Then the realization involves the provision of resources in general.

Needless to say, the thesis of "creation by displacement" has been controversial, even if there is some truth in multiple start-ups caused by the dismissal of waves: but it was able to verify that these companies (crafts, local shops, etc.), created by "hunted" rather than by "hunters" of the productive system, were less efficient and less viable than those resulting from a deliberate and proactive commitment to creation. Shapero also has the merit of showing that entrepreneurship is a repetitive act: often, the creative entrepreneur fully succeeds in its draft after several attempts, in a kind of "pedagogy of errors."

(2) Second, the entrepreneurship action is not limited to a single building project. It tends more and more to distinguish the project design phase of the start-up phase itself. The project design lead to a "layout" resulting in "implementation." Too often, aid systems stop at the project design (existence of a market, reliability of technology, guarantees, paperwork, and various guarantees). However,

the expected and unexpected difficulties start after beginning (launch of the first series, first orders). The main challenges are: start-up capital (working capital) being inadequate (you have to pay suppliers cash and customers pay later), resulting in swelling of the short-term, and product being very expensive; unforeseen administrative difficulties (permissions); technical difficulties (startup problem, technical problems); commercial difficulties (customer requirements, prospecting difficulties); and human problems (difficulty in finding skilled workers, conflicts between partners).

Most often, these starting problems were not anticipated, and, when they occur, the entrepreneur is isolated; but support should occur at that time, perhaps more than when editing the project. The creator will come out unharmed if it initially adopted a comprehensive operational approach and ensured consistency of his plan when developing the business plan.

(3) Third, we must consider the circumstances and terms of creation. There are many cases where the creator is "accompanied." In the incubator, the creator has relayed facilities (three years), various services, and especially aid and advice in the start-up phase. In business parks, it has mostly pooled services (communication, catering, etc.). It is the same technology parks, which welcomed new companies specializing in high-tech sector (in theory). A company can also swarm: an employee may decide to create his own business. He enjoys using his company, which agrees to resume on failure. In fact, we must distinguish the "spread inno-vator" (which often remains linked to the company sponsoring spin-offs) and "swarmed prompted from" mostly resettled in low-technology activities, and detached from the sponsoring spin-offs society.

A company can entice one of his executives to create a new activity (new product, new technology, and new market) within the company, and to support its development: this is what we call intra-contractorships. Generally, these cases are infrequent, though much publicized (the most famous example is the "Post-it" in the 3M company). In fact, "intra-contractors" do not have all the functions of a "real" owner-manager contractor (in particular, the complete autonomy in operational decisions and taking risks to its equity). There are also a number of pseudo-SMEs created or taken over by large groups. These enterprises have subsidiaries that specialise in prospecting and buying out interesting SMEs that are likely to be taken over (either because they are experiencing problems of uncontrolled growth or because the owner is keen to resell it). This case also applies to "small groups" (hypo-groups), where, by the establishment or acquisition, an owner-manager develops on various activities.

In all these cases, the company creating the conditions is obviously more favorable, due to the effect of experience and "amenities" that are offered to entrepreneurs. Thus, in general, the issue of entrepreneurship has significantly distanced us with certainties and highly formalized procedures: it is more a question of dominant psychological process as to when to create a company that procedures based on mathematical logic techniques. Nowadays, entrepreneurial analysis can no longer ignore this dimension of the problems. This double dimension, procedural and processual finds its fulfillment it addresses when the operational analysis problems.

CHAPTER 2

Business Optimization

After a few years down the line, the most important lessons in business optimization focus on the difficulty of integration. The students generally struggle on the case of management of small businesses involving few employees.

Such cases were investigated, which seemed so simple that most participants found it difficult to motivate themselves. It described a jumble industry, the company history, the expression of the characteristics of the company, in particular, its statement of "corporate purposes," and operational marketing approach, and it replicated a discussion between the author of this case and the president of the company. With such a general portrait, students often ask what was expected of them. But these cases were individual companies and, in this presentation, they seemed already unique because of their ability to define and especially to take big decisions consistent with this definition. Gradually, students have been introduced to new aspects of the management of these companies.

These cases already put students to the test: these enterprises could seize the opportunity to buy a competitor. Should they do? Another case revealed the different perspectives of four functional departments. It was discovered that each service has a different mission, different methods, and operational problems. More importantly, every director had his personal philosophy and a different management approach. Some cases showed these managers in action, during meetings of decision, and suggested the difficulties when they might have to act together.

Finally, the last case showed the president who, as responsible for the coordination of this set, suddenly became more complex and delicate, and had to make decisions that could decide the company's failure.

While studying these cases, the students understand the great diversity and complexity of the issues the leaders are facing: market problems (understanding customer needs, competitors' actions), operational problems (to operate plants), the management and leadership issues, issues of power and motivation of staff influence unexpectedly on the operation of the business, the problems that the company's size and the apparent simplicity of its activities that are not announced. How then to bring order in such situations?

In the conceptualization of the general management activities, one of the most outstanding works is that of Barnard (1938) on the functions of the leader. Barnard had first suggested that organizations were "cooperative systems," Cooperation "conscious, deliberate, with purpose" could help people to achieve goals that would otherwise be impossible. All the talent of leadership then was to instill and maintain people's willingness to cooperate. For Barnard, achieving cooperation from people associated with the organization meant that the objectives are clear and appropriate systems "of material stimulation and persuasion" are in place. This had to be done so that there is "a balance between the contributions of those involved and the compensation they receive." The person who agrees to cooperate judges this balance. Accordingly, the art of management is to convince the people associated with the organization that the current balance is acceptable and justified for the continuation of cooperation.

Herbert Simon (1945) has, in a sense, operationalized research on cooperation. He suggested that the unit action should be the decision. The leader then is the one who works to influence the decisions of his/her employees in order to make them converge toward a common goal. This influence, which is somehow equivalent to the effort of maintaining cooperation, aims to act on the factors that may affect the understanding of the goals and their achievement, such as habits, reflexes, know-how, values, and attitudes. For this, tools such as training (to increase know-how), communication (to make clear objectives), and authority (to enforce the people the effects of decisions taken elsewhere in the organization) can be used.

Selznick (1957) studied the reasons that allowed the Communist parties of Eastern Europe after the war to survive a particularly aggressive adversity. It resulted in the emergence a conception of leadership that has greatly influenced the operational approach of organizations. His book, *Leadership in Administration*, suggested that all organizations do not have the same nature. There are those that are simple instruments, putting into practice a technique or procedure, and those who have a "personality." It is the institutions. They are "infused with values," and this gives them a special ability to order dull in-conflict and adapt to disturbances in their environment.

Leaders play a critical role in the "institutionalization" and in maintaining the "character" of the organization. In particular, they must ensure that the values are transmitted within the organization and that the "elites" who are carriers of these values are formed and protect the organization from ex-dull influences. These values form the core of the organization and are the elements of its "distinctive competence."

Christensen, Andrews, and Bower (1973) described the task of the CEO or general manager under three main roles: (1) architect the purpose of the organization; (2) organizational leader; (3) staff leader.

The leader is the guardian of the company's goals. To do this, he presides over the establishment of objectives and the allocation of resources, carries out or ratifies the choice among different operational solutions, and represents the goals of the organization against external attacks and internal erosion. He must ensure not only the maintenance of the organization but also its creative development to achieve the desired performance. The most crucial qualities required of a leader are the ability to conceptualize the purpose and the ability to transmit to the members of the organization. The leader is the main communicator of the purpose of the organization. The leader must also be a careful project manager whose role goes beyond the focus on the goals. He must constantly worry about the integration of specialized functions that tend to proliferate and lead the organization in all directions.

The leader must finally act as a motivator and as a negotiator. He must command respect and be able to elicit cooperation from his subordinates. In areas where the judgment cannot be replaced by detailed procedures, it is often by his behavior the leader clarifies the expectations of the members of the organization. Motivating managers and then assessing their performance are important functions, but are often difficult to reconcile: the first requires an understanding of the needs of people, while the second is based on an objective assessment of technical requirements required by the task. Although his/her function is involved and he/she requires qualities of all kinds, a leader is not a superman. That is when it makes sense to focus on the operational approach to help with such a daunting task.

For Henderson, the first element of Hippocrates method is hard, persistent, unremitting labor in the sick room, not in the library. The second element of that method is accurate observation of things and events, select, guided by judgment born of familiarity and experience, the salient and the recurrent phenomena, and their classification and methodical exploitation. The third element of that method is the judicious building of a theory—not a philosophical theory, nor a great effort of the imagination, nor a quasi-religious dogma, a walking stick useful to help on the Way and the use thereof.

When Hippocrates illustrated his method, he described the doctor's situation had to act despite a lack of knowledge, and despite many uncertainties about cause and effect. For the physician Hippocrates to make decisions, he needed "a familiarity in-first time, intuitive with things, then a systematic knowledge of these things and finally an approach to thinking about it." The parallel with the manager is not only relevant but also striking.

The manager is in a position similar to that of Hippocrates. For students considering the case under review, it was obvious that we needed a tool to bring order, a kind of blind stick to find the way through the clutter of daily management. And today's situation of the companies is even more difficult to envisage the past. What characterizes the management, with the explosion of information technology and the gradual fall of customs barriers between nations is the incredible complexity of contexts and phenomena. Neither small business nor enterprise produce single product beyond that. The complexity not saving any organization, therefore, makes sense of reality that becomes an important necessity for the manager.

In practice, successful managers demonstrate an ability to understand, often intuitively, and create solution that is impressive. Consider two examples reported by the trade press. These are companies that operated for years. A small company that is a subsidiary of a large group was about to close. She had been paralyzed by a strike that lasted 172 days. The Chief, after many failed attempts, succeeded in obtaining a bank financing to buy the subsidiary. Convinced that the company's success depended on employee commitment, the company president wanted to arouse in them a behavior of owners. He not only gave them a portion of the property, now to do this way is rather common, but also decided to teach everyone from sweeper to milling, things that bankers know. During the following year, the company spent hundreds of millions of dollars for training in finance, six times more than for the improvement of production skills. Each week, the company stopped its machines for half an hour to allow its employees to discuss in small groups the last financial documents of the company. Also, a few million dollars were also distributed in performance bonuses.

This policy has paid off for the company, because it is part of an industry where margins are minimal; any attention to costs from employees therefore makes all the difference. More than 1600 companies, including prestigious companies, sent people home to learn from this remarkable experience. Now consider the opposite situation that of the production company. This company was a manufacturing and marketing products company. Its main products were different but marginal products. The company was the result of a merger between two companies. The main competitors were an American multinational, and a firm, the local sales of which were about twice. Years later, the management asked a consulting firm to conduct a study of its competitive business context. The work, very detailed and very accurate, highlighted, among others, the following:

- In its specialty, the company was in a strong position. There were few large competitors and none had a truly national dimension, probably because of conservation problems in long-term products. The company even had a significant know-how that could be exploited in a national and international expansion.
- In terms of ancillary products, the situation was totally different. First, there were strong and vigorous competitors. In particular, the existence of a dominant competitor with a market share almost four times that of the company and spending authority for marketing and product development which was much higher. The conclusion of the study was that in this industry, success was conditioned by traditional economies of scale factors and financial power. Ideally, the company would be increasing its market share by becoming a national player rather than regional, or differentiate by focusing on one type of product.

The corporate culture was dominated by a company with a tradition of quality and craftsmanship that took importance. This culture was also marked by a cost-reduction culture and research of operational efficiency. The most developed part of the study suggested the company to focus on its most promising products and try to dominate the industry. This meant including acquisitions and targeted divestments.

The leaders of the company then began to realize that part of the recommendation, which appeared to them the most obvious and apparently the easiest to do. They made a major acquisition in their area and developed a dynamic market optimization method to remove market share to competition. The only element that was overlooked was the competitive responsiveness. This reaction was devastating. Their products were better known and appreciated by the general public. By combining advertising and price reductions, competition has placed the company in a catastrophic situation. Increasing costs, combined with the drop in income, has created a particularly difficult situation. In addition, the company, after making efforts to carry out a major acquisition internationally by believing in an interesting operation, proved a disaster. The acquired company was facing bankruptcy and asked too many managerial resources to head over the company's operations. The company, at the same time, would have had the opportunity to make a series of acquisitions of quality companies to another of its product segments. This segment was much more consistent with the company's tradition and probably more to the extent of the company, but executives decided otherwise. As a result, the company had to retreat, trying to sell all its activities other than the main segment. Most leaders then left the company.

The leaders of the company did not really follow the rules of Hippocrates. They did not have an intimate understanding of their craft. They made systematic efforts to understand their field, but these efforts were not very concentrated and were dominated by a kind of magical thinking. Finally, they did not have a convincing theory about what animated their field. They are simply allowed to be impressed by the study carried out and tried to implement its recommendations without wondering if they were capable of doing what they were envisioned. Looking back, one realizes that the company was more capable of a transformation that would put emphasis on product quality and differentiation rather than on the costs and economies of scale. It would have been more consistent with their resources and values of employees and managers. The company has since refocused on what it does best and has improved its market position.

Drawing on Roethlisberger (1977), try to recall the analytical framework that Henderson, inspired by Hippocrates, proposed to include social systems such as businesses and to act on them. Here is what he said: you need to have a concept map, necessary to the investigation or understanding, a kind of reference for action; this scheme is not a question of right or wrong, but a question of relevance. In other words, the real test for a conceptual scheme is not whether it is true or false, but whether it is useful and appropriate; this scheme should be used. This is not a theoretical learning object; it can be improved during the action; this pattern is not universal. It can only be used to understand a class of phenomena or act on them. It is a kind of a little primitive instrument, rather than a highly sophisticated instrument; this scheme must be used as long as it provides assistance to those who use them (Beware you think you may have a real bear by a real tail). Nothing could be further top from the truth. You have just a walking stick to assist you here and now. This walking stick will return someday to that glorious graveyard of abandoned working assumptions.

The manager must be prepared for the day when a different way of thinking will be more useful: Commit yourself to a spot of view. Without such a commitment, nothing productive results. Purpose someday that your commitment (your conceptual scheme) will have done ICT job. Be prepared for that day. Be thankful for what it has done. But when that day comes, be of stout heart rejoice and abandon it to God.

Applied Management: this framework is called the business optimization. These companies, like all organizations, sometimes act with a conscious conceptual framework, and sometimes are less aware. The difference between success and failure of an organization often comes from the clarity of the instrument and its relevance. Many organizations fail because they have not been able to renew their approaches when it was no longer useful. In other words, the optimization is a method of action. Business leaders need to discover or find their way in the darkness of an uncertain and turbulent world. But, this is not a universal tool. It must first be adapted to the situation. However, this can never be adapted forever. Be prepared to give up and replace it when it is no longer relevant. Of course, there is a variety of methods, and it happens that a company can adopt a sophisticated method that will prove useless or dangerous to the unwary buyer. As we shall see later, this risk is reduced when the leaders themselves are working to identify the approach of their organization. Companies that are successful usually have worked hard to design their method.

CHAPTER 3

Approaches for Applied Management

The systematic use of applied management is relatively new. In fact, the applied management undergoes various influences that reveal the diversity of its origins and its contributions. Often in business schools, this is entrusted to a corporate practitioner (a professional), which mainly proposed revenue. Conversely, in the universities, it is given to microeconomic specialists and is often in the exhibition very theoretical models or he was entrusted to teachers of management techniques that focus on the technical planning and control, rather than on strategic thinking. Finally, professionals are especially interested in policy issues, to concrete decisions.

It will be understood that the domain is, par excellence, the place of the confrontation of thought and action. It requires an understanding of concepts and ideas, the reading remains commonplace as they do not face a practice, for example, by the case, by his own business experience. This symbiosis perfectly appears in North American education. At first, the authorities are concerned about the conceptual poverty seminars. Because these courses are devolved to practitioners, most often. Universities, to maintain their competitive position (as they are the subject of rankings, which justify the registration fees) then recruit researchers from disciplines "harder" as the industrial economy; these researchers propose more formal approaches.

At this point, it must be said that the definitions proposed in the outstanding works reveal the absence of complete consensus in the current state of the discipline. However, we can classify these definitions around some recurring themes:

- The theme aims: Any approach based on the definition of long-term goals and determining how to achieve them. This type of definition is more interested in policy issues of the company or organization.

- The theme of the plan: Any approach based on a planning commitment of resources over a given horizon. No plan, no operational approach in this extreme design. The applied management then identifies the optimization of planning.
- The theme of the business context: Any decision that aims to make the company competitive in the long term to strengthen compared with a business context where competition reigns. The applied management then identifies with the struggle on the markets and approaches to marketing optimization.
- The theme of change: All decisions involving significant changes, structural, in the management of the company (its goals, activities, organization, etc.).

Often these various meanings are grouped in a very banal formulation of the type: The optimized management is to plan for change in order to adapt the resources of the organization with the requirements of the competitive business context, to achieve the objectives and fundamental goals. It is true that most textbooks are designed with this in mind. We first define the goals and general policy, then we set the diagnostic features of the context of business and the organization before implementing planning ways to achieve operational procedures of activities including performance will be controlled.

However, we see that there are two distinct levels principle: The level of "corporate optimization," which broadly corresponds to what is called the general policy and the level of "business optimization" elaborated in operational products and markets divisions. Of course, these two levels are closely related (in small businesses, they are perfectly combined). But they correspond to distinct problems, including how decisions are made as to their purpose. Now, according to the importance given to each of these levels as to each of the dominant themes, the operational approach of schools of thought are emerging.

Meanwhile, experts have identified what is to define the concept is a type of action consciously wanted, a type of action formalized and structured, an action intended to achieve a specific objective (it is those tactics), or looking for a favorable location in the context of business, to sustainably compete, a perception of the position in the future. These five points are interconnected and underscore the strong link between reflection and action in this area.

Thus, in large bureaucratic organizations, writing the plan plays a key role in the optimization process. The plan will be formalized, run through maneuvers unfold over time and cause some market positioning. In small companies, the process is different. The maneuvers are important (the approach is reactive), optimization emerges from these maneuvers: it is structured on the job, gradually acquires a certain period, some perspective of time, and helps position the company without there being a formal plan initially. The approach to optimized management is the group consisting of thoughts, decisions, actions designed to determine the overall goals.

This concise definition cannot hide differences of design, which can be explained primarily by the diversity of disciplines that have helped forge the applied management of companies and organizations. As we have said, the conception of education is strongly linked to the culture of the teacher: witness the diversity of the content of textbooks. This diversity is explained both by the youth of the discipline, as an object of study, and by the age of the practice. It is, therefore, to trace the sources and a critical review of these.

The relationship is simple, officials in Athens were responsible for the conduct of the war, under the watchful eye of the archons, notables responsible for managing the city (we thus see a first representation of the distinction between policy and optimization activities). It was not until the Napoleonic wars that theorists beyond mere conduct battles to look into the art of war. Karl von Clausewitz, Napoleon observing campaigns, broadens the debate by showing that war is only one form of foreign policy, the diplomacy of a country, violent form, subject to further political goals.

During that time, large US companies sought for thought to develop their management and believed in finding in military theories on the conduct of wars, campaigns, and battles. We live and bloom of many books on the art of war applied to business affinity is much debate. From these discussions, we can draw the following observations: At the simplest level, the warrior image gives rise to many expressions without real depth reflection. At another level, some similarities can be observed between the conduct of business and the battles. First the relationship between the respective forces (organizational resources, competitive advantage), the state of the ground, and terms of engagement (competitive positioning); then, on the conditions of how the battle or war (tactical maneuvers). However, the fundamental objection is that in the business war, this is generally not destroying the competitor (market forces are in charge).

The affinities are actually much stronger in the following two cases: First, when the company's approach is a knockout approach of competitors or the context of business is strongly hostile; secondly, when the military war is not to the destruction of the opponent or when the battles are conducted without seeking total victory (the Gulf War, guerrillas in provide many examples).

In such a configuration, the economic analysis (particularly the microeconomic analysis of markets) provides a low spot in business optimization. Focusing primarily to general equilibrium, partial equilibrium outcome (in each market), the economic analysis assumes that for optimal performance, the maximum profit, the company must just obey blindly to market that are the price signals (wage rates, interest rates, profit rates, product prices). The business owner has to settle for the best use of its resources if it is rational.

This analysis has long prevailed. Today, economists who are interested in business and industry are placing increasing site or critical, the management approach, developed to provide essential tools and models. The main amendments to the traditional economic theory are: There are theoretical situations that are different from the pure and perfect competition and allow the company to choose the amount torque/optimal price. Such is the case of the following situations: Monopoly (single), duopoly (two), oligopoly (few). The optimum can be obtained by different routes, as there is confrontation (conflict), tacit understanding (collusion) or explicit agreement (cooperation). It is not even sure we can logically determine the optimum result.

The more realistic theoretical situation is undoubtedly that of imperfect and monopolistic competition: each company is looking to have a stable market share and adopt an operational approach to survival, not out of war, with too uncertainty. The idea that companies seek to maximize their profit by allocating their resources optimally is a view of the mind; they adopt rather a satisfaction behavior of realistic goals. It is not true that the rate of profit in an industry is only determined by the structures of this industry. It must reflect operational approach businesses that contribute to changing structures (application, technology, etc.) industry in this light, we prefer to speak of Industrial Organization instead of Industrial Economics.

The traditional economic analysis gives no role to the company and the entrepreneur. Joseph Schumpeter will show the key role of the entrepreneur in capitalism, by its innovative approach. Similarly, Coase will show that trade can take place either in a market or in an organization: that internal transactions are fewer "costly" in the broad sense as external transactions, justifies the existence of the firm in a market economy. After a very important current business, economics will focus on the theory of the firm, paying increasing attention to the relationship between structures and operational approach, at the company as the industry. The contribution of the economic analysis, as amended, was primarily to give more rigor to the presentations on optimized management, by clarifying the scope of certain concepts (e.g. diversification or goal setting).

In another form of literature, economic analysis specifically focuses on the following points: The belief in the primacy of the market economy, free competition, as performance selectors, and accordingly, the director role of profit. The interest in a rational approach, methodical, in the operational analysis (diagnosis, location of the problem, choice from rational criteria, the best solution, implementation, monitoring results). The use of analytics within the substantive logic, mathematical logic, as an aid to a business decision. Admittedly, historical analysis is to follow or reconstruct from documents the evolution of the optimization process (key decisions determining changes) followed by a company or group of companies. The goal is twofold:

- Try to identify the "laws" or trends. Thus, Chandler suggested that major structural changes occurred in corporate America were caused by a change management approach in the selection of products and markets. Big business, according to this author, opposes the "Invisible Hand" of market laws, the "Main Visible" of the organization (internal transaction costs) that shapes according to its operational choices.
- Observing the evolution of techniques and management principles, often assuming that the success of a business is due to the adoption of principles "modern" or "advanced" techniques should be transposed to other companies. For example, Peters and Waterman, watching your top performers, list the "key" to their success. Unfortunately, a few years later, most of them had gone downhill.

The fact remains that the observation of business management approaches, even at the immediate history of the daily news, is an

inexhaustible source of information. Regular reading of articles on business life offers constant applications of concepts stated in textbooks and
other books on management optimization. If we abandon the assumption,
often advanced by economists that the optimization of the company is
determined solely by market forces and competition, we are led to attach
great importance to the role of individuals and organizational structures in
applied management. In particular, the choice will bring to power relations
or at least to interpersonal or intergroup relations.

The sociology of organizations has developed strongly after the
Second World War. The main contribution lies in the view called "quota":
Considering that optimization is not determined a priori but it results from
the interplay of forces and events that influence the choice individually or
globally. Psychology also plays an increasing role, since it allows to better
understand the process of decision-making. She is interested in ways that
"know" makers (cognitive process), they "learn" (the learning process),
and they "choose" (decision-making). Strange as it may seem, the relationship is not so obvious. Indeed, the management of a company aims to use
the best resources available to the company; the manager relies on well-
established techniques, in general, and these are the subject of teaching,
of a transmission simple enough (e.g., accounting techniques). Optimized
management, we encounter complex situations, to "ill-structured" problems for which the answer is not always technically possible. As can be
seen with management students, awareness of the optimization approach,
once taught traditional models and the observed practices, may face strong
resistance. Indeed, it is requested to adopt a "cottage" attitude.

In fact, things are moving in the direction of an increasing role of
an attitude "optimizing with" in various areas of management. Indeed,
management techniques have become, in many areas, highly programed,
so that the computer can "make the decision." Therefore, the manager
must now worry more risky decisions, most complex, a programable bit.
Thus, the accountant will increasingly have management consulting, the
chief of staff will make less pay and recruitment, etc. This means that
we look more and more difficult to programable decisions having effects
outside the function, requiring a different mindset (and we do not seek
"the" solution, but "one" solution this is very confusing for the student).

The relationship between management and optimization is, of course,
stronger, so that the terms are often confused. The management has its
origins in the stewardship function of the company. But we can blame this

assimilation to confine optimized management in the internal aspect to the company, implementation of a determined approach from the outside. Marketing, which has the merit of opening the reasoning vis-à-vis the role of the market and the needs of satisfaction. The number of models are actually borrowed from marketing optimization (such as portfolio matrices). However, the approach goes well beyond the marketing alone: thus, at the "business optimization" must also take account of technological mechanisms. One could cite other influences (engineering, political science). What has been said is sufficient to emphasize the extreme diversity of influences. Now, researchers are themselves of different scientific backgrounds and their work shows that diversity. It follows several schools of thought.

The various current trends in business optimization are located on a clock as "the time ahead," we would currently emphasize on decision-making procedures, model-based and proven technologies near schools or more contingent methodologies to lead to interesting approaches to the processes of decision-making in organizations and individuals. The ideal, far from being achieved, would lead to an approach integrating all of these concerns, procedures, and processes, in a "configuration" encompassing all these problems. From there it is possible to locate in time and space, the origins of applied management. The essential idea is that the goals are set by the owners, and implemented by the leaders after examination of the internal and external situation through optimized management. This approach is very logical, it is now criticized for being too much, not to integrate behaviors and risks, to insist more on procedures than on processes, not understanding the implementation issues. Therefore, another approach will address the problems, that is to say a planned management tool for design and business development modes, through the activities, analyzed technologies, products, and markets. It will be observed that the approach is still highly formalized: Gold, was criticized for its rigidity, difficulty to adapt to sudden changes to change approach. Hence profound changes.

Following heavy disruptions in global capitalism, the problem of competitiveness has arisen critically. Traditional activities are no longer the engine of growth (automotive, appliance, consumer goods) and must "invent" new commercial and technological approaches. Hence the development of the analysis of activities in terms of competitiveness (growth potential and profit). The problem is analyzed from two perspectives: First, competitiveness based on specific skills of the company (resource-based approach) and, second, on a relative advantage over competitors, bound a good market positioning (environmental or ecological approach), the two interacting problems. We are then faced with the choice of activities, causing hesitation among some determinism (the positioning causes the performance level) and some contingency ("it depends on several factors"). This contingency appearance takes precedence over mere formalization when discussing business development mechanisms: Analyses become more complex, the choices are strongly relativized, there is a little decisive model.

In this respect, the context of the case is complex and uncertain is commonplace. But this trivial aphorism covers a reality difficult to grasp in the business models. Research has greatly advanced in this area, although it lost momentum to follow the sometimes-brutal transformations "catastrophic" technical, economic, geopolitical, business. The contribution of the industrial economy has been significant. The prevailing feeling is that in this approach, there is no definitive solution for the company. The dynamic optimization variables are predominant and choices need to be revised constantly. This sense of contingency is exacerbated with the analysis of relationships between the mechanisms and organization. It appears as a management system, moving, interactive, subject to constant change; the notion of flexibility is crucial. We can remember that the focus should be on behavior, rather than on procedures, in applied management.

One thing is clear every time: we finally know little about how people make decisions of a strategic nature. Mintzberg and Simon give a key role to intuition. But they show that there are several ways to make a decision, and the process is influenced by many factors. Finally, there is the great forgotten of the analysis: The CEO. This will be the contractor, defined as one that takes important decisions. But the concrete forms of entrepreneurship are extremely diverse, there are particular types of entrepreneurs who have made the subject of much research, particularly related to the explosion of the creative phenomenon of small businesses in the economies developed. Knowing that the mastery of concepts and tools is essential, reflection leads to emphasize the overall look, systemic, integrative, of decision optimization. This must inevitably deal with complex issues, since the phenomena are inextricably, and sometimes inexplicably linked. The analyst should modestly accept a partial ignorance. For educational needs, problems are addressed one after the other. But it must be constantly in mind that each of them is connected to the others. To understand this interaction required, it is customary to use a pattern "diamond," distinguishing the "poles" or "pillars" of the analysis, and then connecting them with two-way arrows to indicate reciprocal relations.

CHAPTER 4

The Foundations of Governance Strategies

The idea of optimizing business decision is very flexible. It is used to describe and understand the behavior of people, those groups, those organizations that they are governments, nonprofit organizations or private companies, and even those countries. The idea of optimizing the decision covers multiple realities. Aspects relating to the "content" of management mechanisms, possible to say what the company did, so that aspects "process" can reveal how he does it. It should also distinguish the "design" of the "formulation" in the management mechanism optimization. It was established two different approaches. The first four dimensions of decision optimization are the definition of product-market, the growth vector synergy, and competitive advantage. The decision optimization can also be designed in four main ways: a plan, a perspective, a position, and a pattern. These different ways of conceiving the business decision are interesting, but the real challenge is to position them relative to each other, taking into account both the context of business, skills, organizations, and role shareholders.

The business decision can first be conceived in the relationships between the company and its business context that environment is considered objective or as the result of the perception of shareholders. The company then tries to take advantage of this environment holds opportunities and when to protect themselves when threatened or perceived as threatening. Optimization is, therefore, a relationship mechanism the context of business, a mechanism to manage this relationship for the benefit of the organization. In managing its relationship to the context of business, the company has to rely on her skills if she wants to make the best of it. To optimize its management applied, it should from its strengths and try to overcome weaknesses.

The company's relationship with the business context is always mediated by shareholders, whether individual or collective shareholders. This is the perspective of those shareholders who give meaning to the context of business and enterprise and, thereafter, presides over the decision. This perspective allows them to establish a link between environment, business, and operational guidance skills. Classical literature in the field is very interested in the perspective of leaders and their values: BI optimization is then seen as a continuation of the leaders. But more recent literature is increasingly interested in the role of other company stakeholders, rather than its own leaders: The perspective of these shareholders is considered important.

The optimization of business decisions may result from different processes. The decision optimization may result from the use of an analytical process formalized before the action takes place. This is called an applied planning process that led to the formulation of an intentional decision. But the decision applied choices can also form during the action. This is called emergent action, groping decision, and actions in daily life. Applied management is, therefore, built through a set of decisions or actions, or both, that are going in one direction and for determining what it is. It is in this sense that the literature speaks of a pattern. To speak BI optimization, it needs to be a pattern that it is the result of decisions before the planned action or it emerges during the action. This is may also be called the driver vein. It is the absence of a pattern (in decisions or actions), which leads to speak of lack of optimization in business.

But the optimization process is intentional or emerging, the mechanism is still interested in the position that the company wants to have may have or should have in its field. Position the company compared with other companies in its field of activity, from the context of business or enterprise skills through a process before action or course of action is what called the decision optimization. The optimization process inevitably leads to clarify the strategic content, that is, to say the kind of relationship that the company wants to meet with his business background.

The pattern—decisions and actions, depending on the consistency and coherence that will be compared with the context of business, capabilities, and business skills, allows the latter to be more or less effective. This performance can be analyzed in terms of acquiring a competitive advantage over other organizations in this field. This competitive advantage can be defined in terms of economic and financial optimization, but it can also be analyzed in terms of social contribution. This is called social optimization. Thus, without claiming to be exhaustive, we can identify several ways to optimize the company: Optimization as managing organization-context relationship; optimization as an extension officer; optimization as an expression of a community of persons; optimization as a driver vein, like building a competitive advantage.

From there, it should be noted that the objective of optimizing the decision always is to position the company as a business context, which context is the competitive environment of the firm, the institutional context or network a company belongs. This positioning can be the result of a deliberate process or process emerge; it can result from reflection of the only leaders or the action of a community of persons; it may be forced primarily by external factors or result more an assessment of internal organizational factors. But in all cases, business optimization allows the company to position itself in relation to the context of business.

As a reminder, in the literature, there are two major trends regarding this relationship. The first trend considers the context of business is crucial and that significantly constrained the company's position. The second stream, recognizing the constraints of the business context, considers that individuals retain a substantial degree of flexibility, they have the ability to choose and even to shape that context. For the vision that has most influenced the design of the links between the binding competitive contexts and positioning of the company, the industry structure determines the optimization of the firm, its positioning, and ultimately performance. But there are other approaches that have examined the constraint that the business context has on the positioning of companies.

The situation of independent firms in some countries is a good example of organizations that are constrained because of the context of business. These companies are always thanking major sector firms, which are both suppliers and competitors. The latter, being vertically integrated, controlling costs and access to basic resources for independents. These are very disadvantaged by the structure of the industry, which favors the big players. They accuse them of collusion and also seek government help to prevent practices detrimental to healthy competition in the industry. Even when the business context strongly constrained the company and its operational choices, this relationship is always influenced by the company, by its capabilities, and resources. Some approaches are more reactive; they focus on the strengths and weaknesses of the company and how its value chain is configured.

Other approaches are more proactive and focus on developing new skills or a reconfiguration of the organization's value chain so that it can compete more effectively industry and to conquer the future. In the latter case, the context of business is considered less binding and the company greater flexibility against the background of the case. Here, note that the company is not a passive entity to face the context of business. It can also be active, both to get the resources it needs to interpret the context of the case and locate them, and in some cases, to create it. This is the case of several family companies where the corporate CEO, made a revolution by deciding to conduct their business in new and different ways. They find that their industries are experiencing a complete transformation. The shops downtown are losing ground to supermarkets, settling in new shopping centers. They also find that the area is in the hands of a few powerful companies. The best approach, they think, then, is to sell many different products to the small group of mass merchants that dominate national markets. The CEO decided to divert the interest of the customer base door products. Already doing business with several companies, they start to buy a range of hardware and household goods manufacturers to expand their product lines. It restructures all its acquisitions to make them as effective as their businesses. The "novelization" soon became the standard for the entire company. However, this form of optimization also has disadvantages: All the weight is on their (customer) side; they can kill you.

In response, the leaders of these companies undertake to make himself indispensable by filling all the needs of their customers. They do not sell a particular item, but, according to a variety of styles and with a variety of prices, all kinds of products. These companies and keep their customers because they offer them is their benefits and that a variety of additional services. We are an organization service. Anyone can make this stuff. It's not high tech. In partnership with customers, companies offer their "tailored" while helping them maintain very low stocks, seamless, thanks to a computerized system that connects directly to the coffers of each store.

This is where the context of the case is not regarded as completely decisive for the company and the role of individual and collective shareholders becomes important. It is then considered that the perspective of these shareholders plays a decisive role in policy and operational choices. Sony Corporation is among the best-known Japanese companies. She participates in a fiercely competitive industry, consumer electronics. In this industry abound genius imitation, such as Casio, Samsung, Sanyo, Toshiba and Matsushita, which produce multitudes of products about standard or, in any case, difficult to distinguish from each other. Yet Sony reigns over this industry as the company most innovative in-warming. Since its creation, Sony has continually introduced high-tech innovations in the wake of each other. Its name is synonymous with transistors, televisions, VCRs, CD players, etc. How did Sony manage to stand out in this way? Many cite the leaders of the company, and its first creator, which has long been the Honorary Chairman of the Board. Policymakers thought, let me tell you my philosophy: The key to success for Sony, and to everything in business, science, and technology for that matter, is never to follow the others. Our basic concept has always been to give new convenience or new methods, or new benefits, to the general public with our technology.

This dual concern "be unique and practical" dominated the culture of Sony. Although on the surface Sony looks like other large firms Japanese catch, it makes things a particular way, in its own way, which reinforces the values of its creators. But the most remarkable in this success story is the incredible influence that the dream of some managers had on the behavior and operation of this great company. Like Sony, all quality companies are dominated by leaders with strong values and beliefs that consider as one of their main tasks to transmit these beliefs and values throughout the company. Beliefs and values are benchmarks that help corporate members to lie in daily action. They allow knowing what is acceptable in the business and what is not. Because of this, they simplify decided by eliminating what is out of the way, the company wants to follow.

The example of large retailers is a good illustration of the difficulty of staying the course. This is the case of those companies that have experienced for nearly a century great success in the distribution. At some point, leaders have tried to redefine the business as a distributor of general products for the general public. Since this included clothing and household products, we established a subsidiary of general merchandise. Unfortunately, the new leaders of these companies have never realized that general merchandise distribution obeyed substantially different from those of food laws. Companies have then survived with regular injections of funds from headquarters. Unable to remedy the situation, the management decided to liquidate these subsidiaries.

In a different but similar way, some companies have recruited new leaders recognized for their skills in the operation of the stock market. They are redefining business as distributors of general merchandise rather than as food retailers. They believed that the expertise in food distribution could be extended to all types of consumer products. Thus, these companies have acquired among other sports equipment retail chains and distribution and sale of general consumer products catalog companies. Unable to manage the business in order to outperform competitors, companies have lost a lot of resources and energy and they had to sell these acquisitions.

Despite these failures, leaders play many roles in companies. They must first be the architects of the purpose of the company. Their task is then to provide leadership to the company. Nothing is more important than ensuring that the company does not drift. Hold the compass, staying the course is essential for the survival of a company. Only vigilance, perseverance, and determination of the leaders can avoid the company of getting lost on lands that are not favorable. However, leaders are to play other important roles. They can be applied to the management of designers, responsible for its formulation and implementation. They can also be the creators of the training context of performance mechanisms.

How leaders are playing these roles is influenced by different individual characteristics. It is in this sense that we said that the decision optimization can be considered an extension of the manager. Literature is full of examples that show this influence. Mention their values and beliefs, age, education, experience and social origin, gender, intellectual approach and the importance they attach to the rational approach, their emotions, their level of cognitive complexity and maturity, degree of liberalism, their attitudes change, their degree of stability, and seniority in the organization, which have an important bearing on their corporate behavior. Honda's story illustrates the potential role of community shareholders in the choice of actions and success of a company. Honda has quickly established itself in the US, as the leader in the motorcycle industry. The company owned two-thirds market share, relegating far behind the competition. This success is partly explained by the fact that the leaders at the summit had some ideas about the company's expansion and were able to make the move from the intervention of vendors and workers of production. If Honda has succeeded is that beyond the creative ability of management, training mechanisms occurred through trial and error, involving all levels of the organization.

This "pluralistic" conception of the company and optimization of business, which reintroduces the responsibility of all stakeholders and their role in the decision-making optimization, is a break with respect to current "elitist" that has marked our designs of the company, applied management and strategic competence that attribute only to leaders. The reintroduction of the community of shareholders in decision optimization models and pluralistic vision that underlies it leads us to look at the daily optimization of action. As for the concept of driver vein (or pattern), we must say that it greatly facilitates the study of optimization mechanisms in companies, especially in large companies and those with a long tradition.

Corporate optimization is the pattern of decisions in a company that determined and reveals objectives and purposes, golden goals, produces the main policies and plans for achieving those goals, and olefins the ranks of business the company is to pursue, the kind of economic and noneconomic contribution it intends to make to the shareholders, employees, customers, and communities. The pattern (or lode driver) is identified from the patterns that can be observed in conscious decisions. But it can also emerge during the action. It is interesting to analyze the decision patterns in each of the major stages of the life of general electric. The firm, there was a time, was developing in all directions, without much order. It was a kind of evangelical expansion. The central idea was that nothing could resist GE if it decided to deal with it. The diversification that followed was unrestrained. But laxity and lack of proportion made that GE has reached its limits. Thus, after having attacked three major markets, or nuclear plants, large turbines and IT, leaders had to retreat to face a difficult financial situation.

With the introduction of planning, the situation has evolved. Previously, planning was largely dominated by economists and experts in operational research. Following the advice of a consulting firm, the leaders then shifted their practices to modern planning methods, including the product portfolio model and the idea of planning as management processes, today elements of their brand widely taught in business schools trade. During this period, they removed 13 product lines (vacuum cleaners, fans, gramophones, pacemakers, etc.). Planning has been strengthened and refined. The dominant aspects were not the market positioning issues but internal operation. Structure issues, management planning and overall coordination systems have grown in importance.

The company then faced another problem that had the size and diversity of the company. To maintain GE's growth rate at a comparable level to that of the gross national product (the only reference valid given the size and diversity of the company) had to be created each year a new industry which size would be about seven billion. These considerable pressures to growth have brought in front of the stage acquisitions inevitable optimal solution. These changes, quickly bring very different behaviors. First, it is a privileged profit rather than sales. The strategic effectiveness (the choice of continuing operations) and operational efficiency (performance in continuing operations) become essential. Thus, the CEO begins to conceptualize the areas in which the company must work, especially the idea of the three circles: The circle of traditional activities, the circle of high-tech activities and the circle of service activities. Outside these circles, it also attacks create a business without borders: After molding GE so decisively, the CEO determined to transform its culture and organization into what he awkwardly describes as a "boundary- less" company. The CEO wants GE to be an enterprise where: (1) internal divisions blur, and everybody works as A-team; (2) suppliers and customers are partners; (3) there is no segregation between foreign and domestic operations and each GE business is just as much at home in South Korea and Paris, France, as it is in South Carolina and Paris, Texas.

Each period we have just sketched is recognizable from an analysis of the decisions taken in the company. When the observation time is long enough, you realize that decisions tend to converge into relatively homogeneous and consistent sets, one can call patterns. These correspond to periods of success when these decisions enable the company to develop or strengthen its competitive advantage. Also, systematic work of the shareholders and the company leads to the construction of a benefit that allows the firm to survive in a context where competition is often fierce. In this context, it is not that the economic and financial performance that ensures the survival of businesses; it may also depend on the social contribution of the company.

The development of a competitive advantage can take forms quite unexpected, as was the case for many companies in recent decades. It has the advantage that can provide a revolutionary technological breakthrough, such as Polaroid; however, we often have difficulty perceiving the many fewer spectacular creations that one, but are the source of competitive advantage for successful businesses. Consider two examples.

The first example is that of Printing. The owner had bought his father a printer located away from major urban centers. The company's products were diversified and the company was in the red. The owner thought that he must concentrate its activities and prevent the segments where the competition was too strong. He had the choice between the book and the press. As the press demanded short delivery, that the location of the factory-made problematic and as he saw the book as a noble product, he decided to focus on the printing of books.

To succeed in the printing segment, we had to be able to ensure high-quality printing at a reasonable cost. Customers (typically publishers) also enjoyed a rigorous production planning so that the agreed delivery deadlines are scrupulously respected and that they can harmonize their marketing programs accordingly. The printing had an advantage because of its location. Labor costs were low and we could get more employee loyalty to the company. This facilitated the formation and allowed to increase quality. To reinforce this, the CEO has renewed its facilities with technologically superior and more efficient machines. He created attractive working conditions.

To satisfy customers, the entrepreneur has introduced new practices in the industry. He recruited salespeople, especially as the industry worked with representatives. These vendors were advising clients on possible printing forms and conditions that were attached to them. As had also computerized the entire production and sales, the vendors were using their portable computer, provide detailed information on delivery times and prices. Orders that were made came directly into the operating system and were processed immediately for production scheduling.

Having captured a significant market share locally, the company managed to establish itself nationally and reproduce the fabric of success. She became a model company whose success has been constant. In the end, the printing was acquired by a multinational. The second example is a travel agency. Like all travel agencies, the company has faced upheaval in the dynamics of the industry, including the massive introduction of direct services through the internet. While all competitors were reluctant, the company embarked on optimizing online (online) very promising. She has worked to build a profitable niche by combining a website high with the strengths of traditional agencies: The human touch. But for cruises, the site is incomparable. You can search by region, destination, and cruise lines. One can also consider the boats with their detailed plans and whether it is a restaurant or a balcony to the desired cabin. More importantly, the company has innovated by introducing the ability to have instant access to a specialist for advice. One can also send questions by e-mail with the assurance of an answer within the next twenty minutes. With this approach, the company has been very successful.

As we mentioned earlier and as we have seen with these two examples, building a competitive advantage requires effort if we want to success-fully stand out from the competition is differentiating itself by the quality of products and services offered either by producing these products and services at a lower cost than its competitors. The first step is to raise the issue "who is the customer?" The actual customer and the potential customer? Where is he? How does he buy? How can he be attained? The next issue is "what does the customer buy?" Finally, there is the most difficult question, "what does the customer value consider? What does he look for When He buys the product?"

To define the activities, it is then proposed an approach that can be summarized as follows: This involves finding out things' oven. The first is the market potential and market trend. How can we expect the broad market for our business to be in 5 or 10 years—assuming no basic exchange in market structure or technology? And what are the factors that will deter-mine this development? Second, what exchange in market structure is to be expected as the result of economic developments, gold exchange in fashion taste, gold moves by the competition? Third, what innovations will change the customer wants, create new ones, extinguish old ones, and create new ways of satisfying his wants. His concepts of the exchange value of gold makes it feasible to give greater him gain satisfaction?

Finally, meeting the needs of customers may lead to a particular configuration of the company's activities with the objectives of strength-ening the activities that are critical to customer satisfaction and reduce the importance of those that are not. It is this effort that allows either reducing costs or improving quality. It was proposed the concept of the value chain as an analytical framework to allow the search for the most favorable configuration. The development of competitive advantage requires a clear understanding of the business functions, and their relationships. It also requires a clear understanding of customer value chains, suppliers, and key competitors. This can be a significant investment, but when the stakes are high.

Another way to consider optimizing the competitive advantage is to consider the resources. An important tool to face the future lies in the set of resources—human, financial, and material—or the possibilities of access to those resources. However, these resources are only useful if the company knows the best use of, which is to say if it was able to develop the skills to operate in organizational processes. Among the most important skills include expertise in management and innovation, know-how in the most important functions of the organization and expertise in the assessment and modification of the rules of the game the industry in which we work.

Finally, optimizing the competitive advantage implies integration of activities that reflects different logics of corporate level, the business unit, and the different functions. Thus, the corporate level is concerned with issues of resource allocation and reconciliation between short-term profits and long-term health of the company. As for the business unit, it is concerned with positioning among competitors. Finally, in terms of features, we focus on productivity and the contribution of each feature to the overall objective. It is in this sense that one can speak of functions optimization: marketing optimization, financial optimization, production optimization, optimization of human resources management, etc. In this context, we can also say that the optimization function is to update to the way the company values. For example, financial optimization is one that achieves the creation and maintenance of financial flexibility. Similarly, the optimization of human resources is to recruit, develop, and reward the talent the company or the unit needs to achieve its objectives. Through their activities, companies contribute to the society to which they belong. This can take different forms, the most important being the production of different goods and services and contribution to employment and economic empowerment of individuals and regions. This is true for large companies but also the multitude of SMEs that characterizes the economy of our societies.

In addition, all companies do not also contribute to their society. The practice adopted by some companies can get them to contribute very positively to the employment and training of a skilled workforce but it can also cause them to delete multiple jobs, which tends to inflate the number of people unemployed or on welfare. The same is seen through business contribution to the respect and protection of the environment: Some have approaches that are concerned about the sustainability of our resources, while others adopt mechanisms whose effect is the resource depletion and ecosystem destruction.

The problem the Nestle Company has faced is interesting to analyze in this regard. Nestle operates in the sectors of food and pharmaceutical production. It is always concerned about its image and always felt it has a social responsibility toward the developing countries. Thus, it was accustomed, when possible, to produce locally. Yet the sale of infant formulas in the Third World has generated considerable controversy and led to a call for a boycott of all Nestle products. The benefits and risks surrounding the distribution and sale of infant formula in developing countries were difficult at the time to decide. The only thing certain is that the arguments of opponents and their attacks against Nestle questioned the legitimacy of the company's shares. Neither the public relations campaign, neither the prosecution nor the creation of an ethics committee has managed to turn the tide.

Nestle's leaders do not seem aware of the significance of this situation. They probably should have taken into account the fact that infant formula sold in the countries of the third world represented only a tiny percentage of company sales. They should measure the cost to the company of a tarnished corporate image. They should have noticed that the action of opponents, whether it was justified or not, coming to an important resource of the company, namely its social legitimacy. The social contribution of business, resulting from the measures they adopt, has the effect of increasing or decreasing their legitimacy. Legitimacy can be considered as an important and useful resource for achieving corporate objectives.

Many now know that a statement of snoring mission, a press campaign, donations to a charity or sponsorship of cultural events are not enough. They must demonstrate to their employees and stakeholders (stakeholders) for all of their activities here and elsewhere that their social presence is beneficial. This may force them to change some of their goals and some practices, whether an investment in the country of military dictatorship, a production process that generates pollution or wild firing practices of their employees. Legitimizing the company therefore often goes by the need to redefine its identity and system of values and meanings shared by its members. By doing so the company can gain a competitive advantage over other companies of a given field of activity. Nowadays, managers, strategists need to realize that building a competitive advantage derives not only from the ability of a company to compete with others in a particular industry but it also stems from its ability to be considered legitimate within the society.

CHAPTER 5

Principles of Corporate Optimization

The classic mechanism of enterprise optimization is apparently the most logical. This is to prepare CEOs to task definition of the policy applied under the control of the board, which represents the shareholders. The purpose of the optimized policy is simple: it is to maximize shareholder value, which is to maximize profit. Moreover, these large companies are actually groups of companies that manufacture, design, and sell a variety of products in very different areas: optimization practices at each of these markets, products must be consistent with the guidance developed at the branch (the corporate optimization distinction and business optimization).

The best-known model is the model "LCAG." The idea is quite logical: the optimal decision is to formulate the general goals first, to identify the major problems, and to choose the best solution and implement it. Most strategists rely more or less on this canvas formulation → the goal; login → the problem; proposal → alternatives; evaluation → choose → implementation. You should know that the process is facing the following problems: after determining the overall goal, we come up against the multiplicity of objectives; the identification of the key problem and options faces the partial ignorance; the choice of the solution is based on criteria from the financial theory (maximizing the value of the share).

However, these limits do not appear to challenge the principle of the model: it is also explained the "SWOT" analysis, which should lead to the "economic" optimization, that is to say in the selection of products and markets, this seems quite logical, include notably the "vertical" link between the choice of goals and plan, and "horizontal" link between the competitive advantage of the business and competitive positioning in the context of business. These two links largely fuel the great debates on the optimization of companies. The fact remains that this analysis has been adjusted or even more vigorous critics.

Examine immediately the criticism currently made of how the teaching of the discipline has been discussed for a long time. The main criticism comes from those who believe that the fundamental problem is to understand and study how optimal decisions are made, what is the actual process followed by policymakers. In this, they oppose the experts who put forward a very logical approach very Cartesian, of analyzing the problems, referring to the process models to streamline operational choices. On the one hand, some advocate a "gradualist" approach, or "emerging" or "incremental" because they believe that the operational issues need to be addressed constantly in the company; others advocate an approach "rationalist," "procedural" as they believe that operational issues are the subject of deliberate choices, planned and heavily argued. The two approaches are not so irreconcilable as it seems.

Nevertheless, the very rational attitude was developed from this vision. The purpose of the business school is to train leaders of very large companies, to get them used to taking general decisions. The LCAG (Learned, Christensen, Andrews & Guth) model and its "SWOT" version provides a framework on which to support the diagnosis and detection of the problem and considering possible solutions, and finally, the choice of "the" optimal solution. Students have extremely complete cases, mostly large companies or organizations (e.g., hospitals), in which there are all the necessary information. They need to reach a solution within a given period. The academic, in MBA, is limited. This type of education is increasingly challenged, especially by the "highlighting progressive changes." This methodology suggests that optimal decisions are and should be taken rationally, logically. Now, they are taken, even for the most important artisanal and intuitive way. The reasons are quite simple:

- First, the decision maker never has all the necessary and useful information. Sometimes, it's too much, but often it is not enough: for example, on the future development, the intentions or results of competitors. In short, information is limited, which limits rationality.
- Decisions are never linear: you have to "loop back" back, go on assumptions, decisions, given results or events. In particular, decisions cause reactions and changes in the context of business. In short, the process is rather systemic. Also, the role of flair is the experience of directing essential. This is working the right brain

(intuitive), rather than the left (analytical) as an image (also questionable scientifically, etc.).

Second, this methodology is applied to large companies, which largely control their sector of activity, even if they are in intense competition. The environment is given, its structure is stable, and it determines the action of the company if it wants to maximize its profit. In reality, the context is highly unstable and even discontinuous: it is because sudden changes, ruptures have occurred in technology and consumption patterns in industrialized countries. In fact, the model is valid especially for FMCG (fast moving consumer goods) industries where large firms dominate their market—food items (e.g., Nestlé), detergents and cleaners (Procter and Gamble), and so on—usually to a few. This is primarily to gain or maintain market share.

This affects only a small number of companies. The overwhelming majority of optimization decisions are high uncertainty about the context of business actions we cannot be content to plan: we must constantly adapt. Now, the LCAG approach suggests that "stewardship will follow" no problem, it will be enough to plan the implementation with procedures in the organization. Beautiful game was to show that large companies have experienced great difficulties in adapting to operational failures (IBM being in this case, a textbook). In other words, strategic flexibility is incompatible with rational approach when it comes to optimize business.

Third and last, the approach suggests that there is "the" solution, somehow hidden, but we must find through logical reasoning. In reality, the decision maker looking for a solution, as satisfactorily as possible: satisfying for him, since it allows him to move toward its objectives or to realize aspirations; satisfactory to those around her since it leads to "positive" performance. In our view, this criticism is essential to the stage of initiation the optimization of business and companies: the student (often selected on logical skills) expects to find "the" case of the solution, which is not without misunderstandings and frustrations … to the point that it is permissible to ask whether to keep this teaching in MBA. Frustration can also come from entrepreneurs who have used consultants in business optimization: to avoid, consulting firms prefer to use "grid" and "models" that streamline the proposals and … reassure their customers, while integrating into their own training and evaluation procedures of their advisors.

The approach is based on a fundamental belief in market efficiency and competition in a capitalist economy based on private ownership of the means of production. The process of valorization of capital employed in the production is carried out as follows: The financial capital used to acquire resources (material, human, financial, and information) that are managed in a business in the most way efficiently as possible; they allow to offer the markets of goods and services beyond the "normal" return of capital and labor entrepreneur, a surplus profit appears transiently due to innovation, according to the central thesis Schumpeter, increasing capital efficiency. Three characters stand out: the capitalist, the manager and the innovator (the entrepreneur).

In the great capitalist enterprise, it is assumed that the managers are at the service of the capitalists: these are represented by the Board of Directors, which ensures that the duly elected leaders value their capital, seeking to maximize profit. For corporations per share, this amounts to maximize market capitalization, that is to say, the share value and prospects of capital gain on resale: the financial criteria are critical to ensure that the goal is reached. This hypothesis can be considered "heroic." Many authors have questioned the uniqueness and one-sidedness of setting the goal. Specifically, the objections are as follows:

(1) *Profit maximization is not clear*: This is whether there is a profit in the short or long term. Indeed, maximizing short-term profit can lead to underestimate the need for investment, essential for the long-term survival. For example, the company must increase its market share: it must undertake modernization expenses, advertising, training, and so on, which will subsequently pay. A purely financial logic may lead to refuse these expenses, not to displease shareholders, on behalf of the sacrosanct law of the market. The model addresses this problem only through pure financial theory, which assumes perfect knowledge of future profits.

(2) *Profit maximization is not operational*: In the theory of markets, optimization is related to a perfect knowledge of all the facts of the problem. In fact, knowledge is imperfect rationality of decisions is limited and leaders seek satisfactory solutions. Moreover, the choice of target profit rate will be negotiated in the company. Each product/market division will set its own profit targets, and overall

profit will be a result of the members of the organization, in other words, have their say, as shareholders.

(3) *To consider the relationship between ownership and management*: Approaches have shown that optimized decisions belonged to a large majority, to employees' corporate executives: they were not fully controlled by the shareholders, too many absentees and dispersed (the capital is "diluted"). Now, these managers will focus on other goals: growth, monetary income and other ("compensation"), and so on, and to the detriment of maximum profit. This theory, called managerialism, must be seriously qualified:

- The search for the greatest possible profit is more plausible that the owner and the manager are combined, as in small business. However, aspirations are much more complex.
- The leader will look even more the maximum profit that will be tightly controlled by owner-shareholders and that they are sensitive to the value of their capital. The most common cases are:

  ○ The leader is controlled by the family. This may be the case of SMEs, but also very big business because the family capitalism is still very much alive.
  ○ Capital is controlled by a block of shareholders, seeking an immediate profit or longer term and intend to judge the leader and his team on its financial performance.
  ○ Capital is subject to violent pressure on the stock exchange by including competitors eager to buy cheaply now: it offers less profit to its shareholders, the less its value, and the more likely a market attack (take-over bid).

The leader will look for even less short-term profit: performance will be assessed on other criteria (growth, technical excellence, social peace, etc.); capital will be diluted in the public; shareholders expect stable income, regular and safe (cases of "dormant shareholders": banks, insurance companies, for large groups, distant heirs to family businesses); principal protected by tricks ("poison pills") or has allies in case of possible attacks stock ("white knights").

In reality, we see that things are very complex; thus, companies go through stages of accumulation, significant investment and valuation, with distribution of profits.

In total identification with the sole aim of profit maximization, considered the ultimate purpose of any capitalist enterprise disregards the concrete process of setting goals within organizations. Moreover, proponents of rationalist approach have sought to integrate other institutions, such as nonprofit organizations. In this approach, the context is seen as an entity made threats and "opportunities" that can be identified on the basis of facts and quantified observations (balance sheets, market share, etc.). More simply, the context of the case is assimilated to the market and competitors. Moreover, the game market structure is widely expected to impose the company limits its operational approach. Critics have been in two directions:

On the one hand, the competitive environment is much more complex. This will be one of Michael's contributions Porter, to show that the industry in which the company operates is subject to multiple competitive pressures, which are not restricted to one set of direct competition. Furthermore, competitive approaches are not limited only to the "cutthroat struggle": companies need stability and often prefer collusion (conflict avoidance) or cooperation. What is more, through their practices, they are shaping their industry structures; different approaches correspond different competitive positioning. In short, the determinism of mechanisms responds, in modern strategic analysis, a more competitive choice contingent vision. Moreover, the sweeping statement that market structures determine the type of competition, and hence the performance of the company is more an ideological conviction than a scientific approach.

On the other hand, we must go beyond the competitive environment and take into account the societal context. In the model of Andrews above, the company is apprehended in terms of values to clarify to what extent they influence the choice of action plans, but after that, the goals and the diagnosis was defined. This corresponds in fact to a libertarian society where economic market laws impose goals regardless of social values. This design has suffered heavy criticism based on the following arguments:

- The values of liberal consumer society were challenged: excessive hedonism and individualism, failure to take account of social concerns (inequality, discrimination) and ecological. This challenge is such that we can talk about a situation of anomie that is to say a challenge to found the industrial society on common values, as revealed the importance of the ecological phenomenon, new attitudes toward family structures, work, national identity, the environment, and so on. These various identity crises challenge the sole purpose of maximizing profits, even if the market ideology has experienced over 80s back into favor (particularly due to the failure of planned economies).

- These values, therefore, must influence the goals of the company. This rehabilitation has taken place through the concept of moral responsibility to the company and what is called the "wave ethics" in the media. But it should be clear of the often confused terms.

In our philosophical dominant system, a moral judgment answers the question of what is "good" or "bad," "fair," or "unfair" (as esthetic judgment or logical). Ethical behavior is evaluated from these moral criteria: each individual or organization will have its own ethics, obviously influenced by them (each perceive differently what is good or bad, depending on particular society in which he lives, his character, culture). For example, one can make a moral judgment on behaviors in case such as the sale of dangerous products, the copying of software competitors, poaching vendors of competition, and so on. Just as compared to examination fraud, each student has his own ethics, even though he knows that it is immoral.

The new fact is that in liberal ideology, it proclaimed that "Ethics countries" face the questioning of hedonistic values, a company "loyal," "honest," and so on, win customers and will do more profit, which is consistent with the finding that the competitive advantage is based increasingly on services provided by, or product side: better sell batteries really have the duration of use announced—this ethical behavior will pay off eventually. Similarly, it is better to sell "green" products? Finally, the personal ethics can be channeled through a code of ethics, common to a community (company, organization, profession), which requires collective behavior rules. Such is the case of professional orders (which can precisely limit the excesses of competition). These values specific to the company in general, or to the company or to a particular profession, will influence the goals of management.

This match goals from senior management and company owners the question of legitimacy. This can be defined as the reason for the existence of such an undertaking, as a social institution in a given company. This legitimacy rests on foundations called to evolve along with the business and the Company. Given the breaks in the industrial society, we are witnessing challenged legitimacy. Thus, producers of detergents, very legitimized in the consumerism, are strongly implicated in a company concerned about environmental problems. Leaders must then get a message—the philosophy of management—which expresses the values to which the company adheres. This concern is reflected in corporate projects. This communication will also address members of the company. This search for legitimacy is particularly difficult for multinational companies that are located in countries where cultural differences can be very strong, prompting reluctance.

Finally, one last objection is the fact that this approach is a little verbose on implementation conditions of the optimization process. This task is devoted to corporate planners, who determine the objectives that will be assigned to all levels of the company, according to complex procedures. The underlying idea is that the "big" optimization is devolved to officers, the implementation being made operational, with the help and under the control functional.

At one time, it has tended to take a more complex approach, due in particular to the need for greater decentralization of decisions, so that enforcement levels have captured part of the strategic decision that the called "business optimization." Applied management is so widely concerned of the link between the "corporate" and "business optimization."

CHAPTER 6

Operational Context in Applied Management

The context is the framework of the company. It requires officers and forced their actions. However, it never completely determines their strategic choices, since the same context will be perceived and apprehended differently by these leaders. Consider two examples. The wristwatch market was originally dominated by Swiss manufacturers, which had strengthened their reputation and position in alliance with the watchmaker craftsman. The watch was designed as a jewel that required the intervention of a consultant and specialist intermediary. When, in early 1960, the Timex Company has embarked on the watch market, the product it offered was not acceptable to the traditional market for watchmakers and jewelers. They considered the product too cheap and inadequate distribution margin. According to them, Timex was not a jewelry watch.

But Timex had in mind another market that of the new consumer in the postwar, young and dynamic, who considered the watch as an instrument that would simply give time and be strong, reliable, and inexpensive. Timex has discovered and built a new mass distribution channel and developed to make unnecessary traditional watchmakers and jewelers, forcing them to the disappearance or a radically different position. Twenty years later, the Swiss watchmakers, who had suffered a lot of market transformation, initiated by Timex and accelerated by the arrival of Asian competitors, invented the watch fashion accessory. Swatch is not only a reliable and inexpensive watch, but it is also a fashionable watch, that transforms to meet the different needs of new generations of consumers. Timex has been in turn moved.

When Hopp and Plattner launched SAP financial accounting system, they identified a need that nobody perceived. Manufacturing companies, at the time, were concerned about the difficulty of quickly dispose of reliable information on cost of products and processes. The need was widespread.

Consulting companies were generally able to meet the need, but the answers were always custom-built at considerable cost. Hopp and Plattner saw an opportunity: to offer a basic solution that everyone could directly or with external assistance, adapt to its own needs. The automation and simplification of internal accounting operations have been standardized.

Later this program was completed by a series of other modules compatible with the first for purchasing, inventory management, production and verification of invoices, personnel management, and so on, which has become the software package integrated management (enterprise resource planning, or ERP) most popular on the market. One could almost say that when Hopp and Plattner have developed SAP, they perceived a context that no one had seen before. However, since the launch of their first software, the operating environment has changed considerably. First, some competitors are offered products that by returning to the basic modules were capable of being "the best market" such software that took better account of the needs of a particular function.

Thus, PeopleSoft, originally a specialist in human resource management has been able to steadily increase its market share, has had, for example, a sales growth of over 50% at the turn of the millennium. Similarly, positioning itself in the small or medium-sized enterprise (SME) market, JD Edwards took the lead. More importantly, SAP software had been developed at a time when reengineering, conceived as a means to reduce costs, dominated. Today, the search for growth opportunities is more appropriate, and new competitors, such as Calico Systems, Siebel Systems, i2 Technologies and Manugistics, focus on software that can make decisions not only reduce costs but also to increase profits. They take to SAP increasing market share.

This is the importance of the operational context, and analysis tools that enable better leaders to understand this, we will devote this chapter consists of four components. In the first part, we describe the company as an open system. In the second part, we will focus our attention on the competitive environment of the company and the importance of that context for corporate strategy. In the third part, we will show how the operational environment of the company affects the way a company is within its industry and the strategic choices she makes. In the fourth and final installment, we'll talk as network context and the importance of business networks in the current business context.

To understand what the business context is, it is useful to make a detour on organizational theory. The company has long been considered a closed system, with an internal logic fairly immune to external influences. This idea was conveyed by two currents that have greatly influenced our understanding of businesses, see his scientific and administrative organization of work and human relations school. In both cases, we conceived that the company could be effective in correctly applying certain principles of internal operations: specialization of tasks, command units and piecework (in the case of scientific and administrative organization), motivation and worker satisfaction taking into account the informal structure and appropriate leadership style (in the case of the school of human relations). The company was thus conceived as a machine or as an organization, whose dull elements are related. It was enough to be concerned about the state of these dull items to ensure that the company is performing well.

Systems theory contributes in one way quite special and renews our outlook on the company's relations with the operational context. Under the influence of Von Bertalanffy (1968), in the 1930s, systems theory first developed in biology: the body is designed as an open system that interacts with its environment and evolves under the influence of endogenous and exogenous factors. Then the theory of systems fast gaining the field of mathematics with Wiener, who in the late 1940s created a new field that of cybernetics, based on feedback and self-regulation systems. At the same time, Shannon, a telecommunications engineer, published his mathematical theory of communication. Ashby is interested in coupling open systems and, Forrester attempts to apply the theory of systems in industrial dynamics. It is thanks to these contributions in several scientific fields, which gradually develop the so-called system theory.

This view of the world is gradually gaining social sciences from 1960. There will then be a series of contributions, both theoretical and empirical. Two approaches shape the development of the theory. Some authors, such as Talcott Parsons (1960), are felt mostly integrated in society as a system consisting of subsystems economic, political, and cultural and community. Other authors are particularly interested in organizations and the relationships they have with the social system. The company is well regarded as an open system that imports some elements of context, transforms, and exports again in the operational context. This transforms inputs into outputs system, which can be represented as follows:

Katz and Kahn (1966) establish nine common specifications at any open system, so companies considered open systems:

(1) imported energy;
(2) transformation of energy;
(3) exports of goods and services in the operating environment;
(4) the cyclical nature of the exchange of energy in the form of compensation in money or satisfaction;
(5) the possibility of negative entropy, meaning that the organization can store energy and it is not inevitably oriented toward degradation and death;
(6) present information in the form of negative feedback that allows the organization to correct its mistakes and adapt;
(7) a dynamic homeostatic state, that is, a state of quasi-stationary equilibrium that preserves the character of the system will growth and expansion;
(8) a process of differentiation and development whereby more diffuse and global models are being replaced by specialized functions; and
(9) the principle of equifinality by which a system can achieve the same result by following different paths.

This conception of the company as an open system imposes progressively in the field of management and encourages managers to be concerned with both internal dimensions of the company and the relationships it has with its context; these relationships are critical to the overlife and development of the company. But this opening of business on the business context exposes them to constantly deal of uncertainty, which can be debilitating. That is why according to Thompson (1967), companies are open systems and navigating in uncertainty seek by all means to reduce it and developing a strategy can be considered one of those ways.

Systems theory has had a major influence on our view of the company as an open-system context and, by extension, our management philosophy and strategy, because this company is widely accepted design nowadays. She also continues to serve as a classification scheme, useful to categorize and link internal and external variables that are most important to the enterprise development. Other approaches have been developed in sociology and economics. They are particularly interested in the relationship between companies and their operating environment. Some of these approaches are very deterministic. This is the case of population ecology of organizations, studying the sets of organizations and quasi-ecological selection mechanisms that make some Entre taken to survive and others disappear. The operating environment is considered so decisive that it "chooses" the companies that survive, so those that are most appropriate within the meaning of Darwin. Such an approach assumes that leaders have no ability to make choices, and so it is of little interest in strategy.

Other approaches are much less deterministic and allow the executive to maintain its ability to choose and act, even if that ability is constrained by the operating environment. This is the case, among others, the institutional approach (Scott, 2001) and the model of the resource dependence (Pfeffer and Salancik 1978). It is this approach that underlies the analysis we make the operational context and its links with the organization. The leaders when formulating a strategy must consider two types of context: the competitive environment of the firm and the business context. Although these two types of context are interrelated, we will address them sequentially.

Admittedly, any company belongs to an industry that is, somehow, the environment in which it operates. It is therefore important that leaders define well their company affiliation industry. According to Porter (1976) and the movement of the industrial economy, the definition of an industry is from the identification of all groups (suppliers, customers, potential entrants, substitutes) that interact with businesses in competition in a field. This definition is a judgment and so is arbitrary. Nevertheless, this definition is important because it draws the boundaries of an industry. These boundaries are not immutable, because the actions of the company and its competitors, technological and strategic innovations and, in particular, marketing actions help to modify them.

The conventionalist theory (Gomez, 1996) is another method to define the industry. Although this method is promising, it is not yet accurate enough to be easily used by analysts. The analysis of the competitive environment of the company that we offer focuses on three main aspects. First, we present a useful model for the definition of an industry and competitive dynamics. Second, we will address the question of change in an industry and its implications for business strategy. Finally, we will examine the relationship between membership of an industry and company profitability. The structure analysis model and the dynamics of the industry which currently the most used strategy are that of Porter (1980, 1985). Unlike the Andrews Model, which we discussed earlier, that of Porter, from the industrial economy, deals only very few elements of context that are not economic. It also makes it possible to identify key players in an industry and analyze the dynamics of competition that prevails in the industry.

In a text published in 1994 in *The Relevance of a Decade*, Michael Porter summarizes his intellectual and fundamental aspects of its approach to strategy. Influenced both by the work of Andrews, Christensen and based on the general policy of Directors and those of Caves in economics, Porter designs his theory as a synthesis of these two approaches: he wants to retain wealth and multidimensional nature of the cases discussed in general policies of administration and the mathematical and statistical rigor of economic studies. He summarized his theory by seven elements:

- The company must have a clear goal, which is to obtain a high rate of return on long-term investment.
- The strategy is the means used by the company to achieve this higher profitability.
- The unit of analysis for the development of this strategy is not taken Entre but the industry, defined as a group of competitors seeking success with a particular product or service.
- The formulation of a strategy must simultaneously consider two elements: the industry structure and the relative position of the company in the industry. Both are different although several analysts' strategy and economic tended to confuse them, assuming that all industries are similar or that all companies in an industry behave similarly.
- The diagram analysis of the industry structure includes five forces, including new entrants (or potential competitors), the substitute products, customers, suppliers, and competitors. This scheme can be considered an expert system to define the elements that lead to profitability in a particular industry and how these elements interact.
- While the five forces explain the differences in profitability between industries, the positioning of the theory tries to explain the differences in profitability between firms in the same industry.
- The positioning theory has its roots in the concept of sustainable competitive advantage, that is to say, we can maintain over a long period of time; competitive advantage stems from the discovery

and use of unique competitive levers and different from those of competitors. We return to these levers on the internal analysis, on strategic options, when we discuss in detail the generic strategies. We can group the five elements of Porter's model under three themes: the demand for the product or service, the actions of suppliers and the actions of competitors.

A careful examination of the changes in demand for the product or service is an important step of the analysis. In fact, the strategic judgment requires knowledge of the major trends in demand: Can we talk about cycles? Declines in demand, which remains stable or increases? Once we know the evolution of demand, we must seek to understand the reasons that may explain these major trends. In business strategy, mainly two aspects are examined: the behavior of customers and the threat of substitute products.

One cannot understand the request without considering the customer behavior. What is important at the start is to know the customer-base characteristics, for example, their number and their demographics. Some companies may experience a dramatic drop in the number of consumers of their products. In certain regions, companies were making rosaries or missals have experienced this situation due to a large decline in religious practice. Other companies, such as we have seen in speaking of the overall context, must adjust their products and services according to the changing demographics of the population.

Beyond these basic characteristics, strategic thinking requires the company tries to understand customer behavior: what it seeks, the function has the product for it and the characteristics of the product which, in its eyes represent the value. Consumers are becoming more educated; the majority of women work outside home; couples travel and learn about different cultural contexts. All this affects their behavior as consumers.

Overall, the consumer is more sophisticated than it was. The sophistication and the need for individuality accompanying sets, among other things, the phenomenal growth of the fashion design industry and the cosmetics (for men, women, and even children). The consumer is much more critical than it was. His loyalty to a product cannot be taken for granted and he does not hesitate, individually or collectively, to express its discontent and assert his rights.

Sometimes, a company's customers are not people but other companies. In this case, the buyer may have a considerable influence on the one that produces the good or service, especially when it buys a significant percentage of the vendor's revenue. Moreover, it may happen that the power of the client company turns against her. Indeed, if it becomes too demanding, the manufacturer may be tempted to sell itself its products. This is what happened in the PC industry and helped transform it radically. The substitute products also strongly influence demand. If, in a given industry, there is a decrease in demand, it is likely that the product is in decline.

Over the past 10 years in the beer industry, a decline in consumption was observed in certain provinces of Canada. At the same time, there was an increase in white wine consumption. Is this a phenomenon of the substitution of one product with another, or a cap customer beer drinkers and the emergence of a new market for white wine? The analyst must be able to determine. The substitution is only possible when in the market; there are companies that offer different products but perform the same function. This is the case in visual prosthetics industry: the customer has the choice between conventional glasses, contact lenses and in some cases, laser surgery. This is also the case regarding domestic energy: natural gas has become a substitute for electricity. Hydro-Q must deal with the major player that became Gaz Metropolitan; both struggling to gain market share.

If the client realizes that a product can fulfill the same function, another cannot count on his loyalty to the product. Hence, the efforts by companies are to build customer loyalty to guard against not only competitors but also against threatening substitutes. Changes in consumer behavior and the appearance of new products and services are substitutes or competitors that a product or service has a life cycle. Marketers have shown that a product or service evolves through various stages, which they called "stages of the life cycle." Thus, a product or service is entering a phase of strong growth after a period of gestation and introduction during which the demand is stagnant or low growth. This was the case of personal computers from 1975 to 1985. Similarly, the decline in demand can be attributed to aging of the product or service and its replacement with more advanced and more appropriate products and services. The classic corded phone is an example of a product that has entered a mature phase very advanced or declining.

Despite the interest that the concept of life cycle of the product, Porter (1976) rightly insists on the fact that the life cycle should be seen as a consequence of the dynamics that exist in an industry. The life cycle is the result of penetration and then to market saturation, saturation associated with a lack of innovation on the part of incumbents or changes in the size of the buying group. We must therefore focus attention on the fundamentals of the industry dynamics, and not first on the life cycle of the product. Suppliers are major players in an industry. By their actions, they can cause a chain reaction in all other industry players. The most obvious example is probably those oil producers: if they work together to control prices or to regulate the amount of oil on the market, refiners and manufacturers of derivatives should reconsider their strategy as soon as possible. The role of suppliers will be even more important that they are few and they can act together. The reverse is also true: ore suppliers, those raw materials processed and those little amenities have very little influence in many industries.

They align their prices with world prices because they offer a standard product to which they add little value. Refiners and derivatives manufacturers will have to reconsider their strategy as soon as possible.

In general, suppliers have power if the following criteria are met: they are a few (oligopoly); their products are not substitutes; the buyer has no bargaining power (either because the volume purchased or other strategic considerations); their products are important inputs for buyers; they have differentiated products; buyers must incur costs to change if they change their source of supply; they can integrate forward and do what their customers are doing now. For example, a company, such as Group Canam, specializing in the manufacture of metal products, may buy its steel sheets with several steel mills. Once the company has determined its requirements, it can "shop" and choose the provider that offers him the best price. This price will be more or less interesting for Group Canam, according to more or less that have different providers with respect to this company.

But even if the steel sheets are a standard product, the price cannot be the only customer choice element. Steel mills are often distinguished by the proximity of their facilities from those customers and the quality of services they provide, therefore, more generally, for the value they provide to their customers. Sometimes, the power of a supplier turns against him. Take the example of a company like IPL. This company is a supplier of some auto-manufacturers. In circumstances that are favorable to it, it can assert its authority in respect of these manufacturers, but the risk that the automobile manufacturer decided to produce himself before the pieces he bought at IPL.

In the industrial sector, several companies are usually competition for the production of goods and services sought by customers. Every business must know its competitors, that is to say, their characteristics, and the instruments they use to battle. The company must be concerned not only with its current competitors but also potential competitors, that is to say, companies that want to enter the industry. The more the product or service is in the growth stage of its life cycle, the more the industry is attractive to new players. However, there are barriers to entry that prevent or make difficult the arrival of these newcomers. According to Porter and the movement of the industrial economy, there are six main barriers to entry:

(1) *Economies of scale*: In some industries, you have to be able to enter the market with a similar volume to that of competitors; otherwise, we faced necessarily higher costs. Reducing the cost shut uniform is that, on one hand, larger facilities require a smaller investment in the unit and, on the other hand, the more we produce, the more we can, because of the experience, reduce operating costs. Economies of scale exist in almost all industries, except perhaps those that require a substantial and constant adaptation of the product or service to differentiated customer requirements.

(2) *The differentiation and strong branding*: In some industries, product features and brand are elements that determine the purchasing behavior. As usually very expensive to create a brand image or product differentiation by quality or characteristics of use, newcomers are discouraged and avoid those markets. This is the case for the luxury industry (apparel, accessories, perfume, and jewelry).

(3) *Capital investments*: To successfully gain a foothold in certain industries, invest huge sums either to purchase equipment or for research and development. For example, in the pharmaceutical industry, it costs at least $100 million in development costs before selling the first product. The capital required by the investment acts as a barrier to entry.

(4) *Access to production factors*: Sometimes, new businesses have to face disadvantages that have nothing to do with economies of scale. This is the case if firms already active in the industry control patents or access to raw materials or technologies. For example, the Saudi Arabian petrochemical company Sabic has a substantial advantage in the production of major petrochemical intermediates because of its privileged access to the united gas resources.

(5) *Access to distribution channels*: Most consumer products and many industrial products require that routes bring them to access points to customers. If distribution channels are controlled by companies already active in the industry or if access requires expensive initial investment, it will be difficult for a newcomer to find his place. Thus Renault, before the acquisition of Nissan, has never been able to settle permanently in North America because it has, among others, overlooked the importance of distribution channels.

(6) *Regulations*: The need to obtain permits and government authorizations can be a significant barrier to entry. In addition to the difficulty of eligibility for these authorizations, costs, and delays associated with recent reinforces this barrier to entry.

By examining newcomers, we must never forget that they could be coming from abroad. Due to the end of protectionism in most sectors of the opening of markets and globalization in a growing number of industries, new players are becoming more numerous. Besides the effect of the aforementioned factors, the intensity of competition also varies depending on the number of competitors and the strength of the latter. The economic analysis shows that the presence of many competitors in a market that is associated with a lively and open competition. We observe the same in oligopoly situations when the industry experienced periods of great technological and regulatory changes.

This is the case in the global auto industry, a mature industry transformed by major technological changes and globalization of markets, where few big players (following numerous mergers) to better compete position in the markets of Western countries and in the global markets of the countries in transition to a market economy. This is also the case in the telephone industry in Canada, where Bell Canada's monopoly on long distance no longer exists due to deregulation. The competition is now live and it has generated in recent years, significant changes in market shares, even if the prices for consumers have not changed much. However, it was the stabilization of the industry.

In an industry where there are many competitors, a company rarely competes with all the others. In fact, it competes with companies belonging to the same strategic affiliates. Called "strategic group" all companies that, according to some important strategic dimensions in the industry, approaching the market in a similar way. Companies can look through the range (low, medium, or high end) of products and services, the type of distribution channel they use, with their emphasis on customer service, and so on. It is possible to build a graphical representation of these groups, called "map of strategic groups" to build such a card.

For example, if we wanted to make the map of strategic groups in the garment industry, we could use two key variables in this industry, namely, the client referred (man, woman, child) and price. We would place in the same strategic group companies, based on these two variables have the same behavior. Thus, in the area of women's clothing at higher price, we would find not only the big brands of ready-to-wear scratched foreign designers (Saint Laurent Rive Gauche, Donna Karan, Max Mara, etc.), but also some designers (Marie Saint Pierre, Michel Desjardins, etc.). All these designers compete, but they do not compete with companies, such as Peter Nygård or Liz Claiborne.

Some companies do not try to enter an existing strategic group but manage to create a new one. This was the case of the company Harlequin in the book industry. For the production of its romance novels, she decided to operate on completely different rules to those in force in the industry: product standardization, commoditization of the authors, low production costs, and mass distribution. This was also the case for the company, The Body Shop, founded in 1976 in the United Kingdom by Anita Roddick. Favoring natural products based on a philosophy of protecting the operating environment, the company has developed a range of different health and beauty products from those of other companies in the industry. Furthermore, The Body Shop dedicated to packaging and advertising a percentage of the price of these products significantly lower than the others. As the strategies adopted by Harlequin lead to very high profitability for these two companies, we have seen the arrival of new entrants, creating a group with the same basic strategy.

When a company owned by a strategic group whose position relative to the other groups is structurally weakened, its market share tends to decrease. In this case, unless the business is a very important player, able to reposition itself more favorably in the same strategic group, it will study the possibility to change group. Otherwise, it will have to consider redeploying in another industry. This is the strategic rationale made by several owners of independent pharmacies in some countries. Given the importance of market shares that are monopolized large companies chains, sold their pharmacy to larger chains. Similarly, on another scale, the aerospace division of Bombardier was gradually transformed by moving toward an original definition of its industry and a favorable repositioning in commercial aircraft small.

Compared to the old, Canadair, the company completely changed strategy and strategy group. Identifying forces in an industry allows managers to understand the dynamics of the industry in which their company belongs. It also allows them to identify business opportunities available to their business. The company completely changed strategy and strategy group.

Several factors cause structural changes in the industry. Porter (1976) lists eight factors that leaders must consider if they want to predict in which direction their industry is headed and thus make informed strategic choices:

- long-term changes in the growth rate of demand;
- learning that prevails in an industry, both on the part of buyers from the companies about the technology of their competitors or their ways (this is what is represented by the curve experience);
- increasing the size of the market and that of the company, for example, due to globalization and increased market penetration;
- innovations that develop both in industry and outside the industry; For example, customers whose needs change, which require the industry to innovate;
- changes in input costs; for example, the significant increase in energy prices in the 1970s and 1980s that changed all industries;
- changes in the structure adjacent industries. If Microsoft was forced to split into two companies, competition that would result could cause significant changes in products and, therefore, considerable effects on customers and current Microsoft;
- social changes and government influence;

- entry into the business industry evolving in other industries. Thus, the arrival of Philip Morris in the beer industry has allowed a tremendous repositioning for Miller beer and a significant increase in competition in the industry.

We discuss many of these changes when we talk about the company's general context (in its sociocultural dimensions, political, technological, and economic) and its impact on its strategy. However, it seems important, in this section, a little more interest to us to increase the market size and the firm, and the effects of these increases on the company's strategy. As Porter points out, the growth of the market size is usually accompanied by an increase in the size of the leading companies in the industry. Increasing the size of the industry and that the company has effects on the structure of the industry.

First, the increase in the size of the market and enterprise tends to expand the range of possible strategies for the latter and often has the effect of increasing the importance of economies of scale or capital requirements. The example Cessna light aircraft in the industry is revealing. Increasing the size of the market and that Cessna has allowed the company to move from a production unit in mass production, which made possible economies of scale and created a competitive advantage in regard to the cost.

Second, due to the increase of the market, suppliers and buyers increase their sales and purchases and are increasingly tempted by vertical integration strategies. These changes in the industry will inevitably have effects on smaller firms which are unable to have the volumes necessary to take advantage of economies of scale or a vertical integration strategy. They must rethink their strategy director to develop skills that will offset the very low cost of industry leaders and the benefits they receive from vertical integration strategies. They must think about the elements that increase the differentiation of their product compared to those of industry leaders, such as the development of new products.

Third, the growth of the size of the industry usually attracts new entrants, which is this time a particular threat to industry leaders, especially when these new players are of a large size and they have developed transferable skills in the industry in which they want to enter. This occurred in the recreational vehicle industry by the late entry of companies belonging to large agricultural equipment industry. Their entry has forced companies in the industry to rethink their strategy.

All these examples show us that the strategy of a company is strongly influenced by the structure and dynamics of the industry to which it belongs. The industry does not determine the company's strategy, but it frames and forced it significantly. Furthermore, due to its strategic choices, the company also affects the structure and dynamics of the industry. The case of the company—Cardin—illustrates this last statement. Traditionally, the luxury industry has developed through differentiation strategies adopted by all companies in the industry and thanks to the great control that enterprises exerted on both the design of the production and distribution of their products. Pierre Cardin has decided to break with this tradition. It adopted a growth strategy licenses, abandoning all control over the production and distribution of products marketed under the brand—Cardin.

The creator was first put on trial by the industry representatives. But before the financial success of its strategy, they were forced to rethink their own ways. While refusing widespread use, licenses as Cardin was, since their eyes this might kill the quality, brand image, and ultimately the industry, they have considered broadening the range of strategies and are resigned to join big luxury conglomerates. The structure of the industry is thus found significantly changed. Belonging to a particular industry can she explain the very high profitability of some companies? What is the importance of membership in the industry compared to other factors such as positioning? When McGahan (1993, 1999) examines various profitability indices [return on equity (ROE), return on assets (ROA), and return on sales (ROS)] US manufacturing firms, he made three observations: the profitability of manufacturing companies varies from year to year; some manufacturing industries (drinks) are very profitable, while others (iron and steel) are very little; in the same industry, companies are making very different financial performance.

His analysis led the author to the following conclusions. On one hand, some companies, which are nevertheless in highly profitable industries, have low returns on their investments. This means that the profitability of the industry does not guarantee that each of the companies belonging to the industry will have high profitability. On the other hand, some companies belonging to very few profitable industries manage to have high profitability. This means that strategic positioning choices really matter. Finally, differences between profits and profitable businesses in certain industries can be very low, which means that there is a relatively wide range of profitable positions in the same industry.

The research strategy therefore seems to suggest that a company reaches a high threshold of profitability when it belongs to an attractive industry and it manages to keep its competitive advantage. It is the combination of these two elements that combine to generate the greatest profitability. The relative impact of one or the other of these factors, however, varies across industries. The model of Andrews, unlike the models from the industrial economy interested primarily in competitive environment, also attaches importance to the operational context. This model forces the leaders to question the major trends in the context and identify opportunities and threats it contains to have a clearer idea of what the company "can do."

We discuss five types of context: (1) demographic, (2) cultural, (3) policy, (4) technology, and (5) economic. About each, the leader must ask how this element can change the structure and dynamics of the industry, and how it can change its position in the industry. These are the demographics of the general population, namely its age structure and its evolution, its distribution by gender, ethnicity, religious affiliation or level of education. Marketers interested in the behavior of consumers place a high value on these characteristics. Indeed, the ageing of the population, the massive influx of women into the labor market (in economically advanced societies), or the increase in the level of education have significantly changed the characteristics of consumers and their purchasing behavior.

In industries such as clothing, food, tourism, and financial services, to name but a few, sociodemographic changes have prompted companies to make choices that reflect these new realities: clothing "woman career," clothing to new lines but for an aging clientele, educational trips of all kinds, and so on.

The changing demographics strength designs new strategic directions. It may also have an impact on the capacity company to acquire the human resources and technical expertise they need. For a long time, companies are not much concerned about their ability to acquire these resources. First, as business activities have little or no specialized workers and employees, they had access to a large labor they often formed on the job. Then the standardization of tasks and mechanization have made sure that the workers were easily interchangeable.

The advent of the knowledge economy has changed the landscape. Companies leading sectors of the economy, such as telecommunications, multimedia, or the pharmaceutical industry, need employees with extensive training in these areas. These resources are in limited supply in the market. This explains why the labor market of persons holding these skills is so effervescent that companies are in fierce competition to acquire and wages are high.

The aerospace division of Bombardier had this rare resource problem in aeronautical engineering. While its strategy to expand the incited to invest heavily in the aerospace industry, the company had a lot of difficulty finding engineers and aviation technicians she needed. In the short term, she had to have them come from the outside and, in the longer term, she worked on the development of training programs in the field of aeronautics. Conversely, one might think that the development of the multimedia industry in Montreal is linked in part to the presence of abundant labor and specialized in this field.

By cultural background, we hear all the norms, values, beliefs, and ideologies that characterize the society in which business operates. The interest of management science for culture gave birth to the United States to what has been called culture and management. Fritz Rieger is in this stream of thought. He shows us how the national culture affects the strategic decision within five aviation companies he studied. Multinational companies wishing to geographic expansion know they must carefully analyze the cultural context of the countries where they are thinking of moving because certain cultural characteristics of populations may constitute major threats to achieving their strategies. This is the case, for example, companies that are trying to pursue a strategy of "total quality" and "zero defects" in sociocultural contexts in which the concept of quality is new and not understood by the people concerned and the workers of these companies, or which covers dimensions are those typically associated with this concept. But cultural characteristics can also be favorable to the company can exploit them. This is currently the case in Asia in many sectors.

Although culture is slowly changing, it changes inexorably, which explains that individual expectations are changing. Environmental issues are a very good example. While for a long time our companies are very concerned about ecology and respect for nature, they have become very sensitive to these issues. Citizens accept less that destroy forests haphazardly in the name of maintaining employment in a region that pollutes the waterways in the name of maintaining agricultural or industrial activity and polluters are not liable. The citizens of industrialized countries are increasingly concerned about sustainable development, but this concern is also accentuated in developing countries. Placer Dome allows us to illustrate this change in mentality (Sloan, 1999). Placer Dome is headquartered in Canada. Entre had taken a minority stake of 40% in Marcopper Mining Corporation, a Philippine company that operates a copper mine.

In 1996, a break caused major leak residues in the Boac River, which had disastrous consequences for the local population. Placer Dome is a company for a long time in international forums on the protection of the operating environment, and its leaders have long been sensitive to this issue. In 1989, Placer Dome has been clear on its commitment of management to protect the operating environment and the sustainable development of local communities in which it is installed. The ecological disaster of the Philippines was therefore an important test of the seriousness of his intentions. When it became clear that the Filipino shareholders did not intend to release the funds needed to remedy the situation, Placer Dome (although a minority shareholder) agreed to accept responsibility for all costs associated with the cleanup and compensation of populations. The company has even released the funds for a phased program of sustainable development over 10 years. This responsible behavior reflects the profound cultural change taking place in our society in relation to ecology; change is not without influence of many business leaders, such as Placer Dome.

The political context has several aspects, including the political regime, regulation, and taxation. All these can change, significantly, the dynamics of competition that prevails in a particular industry. Business activities are influenced by the type of government in power. Some political systems are "natural allies" of companies, and it can lead to generous grants and business assistance programs (technological catch, export support, training workers, etc.). This is the case of conservative and liberal political parties in Canada. Other political regimes are often considered the "al-related natural" trade unions. This is the case of the Socialist Party in France, of the Labor Party in England. Business leaders know that the ideology of the political party to see importance for the activities of their companies. For example, several governments of industrialized countries regard the regulation as a substitute for Crown corporations. Rather than making himself certain economic activities, the state supervises and directs. This regulation, however, is a limitation and a constraint for the company. Just think of the regulations on logging, which now requires companies with more countries to be reforested, or laws on health and safety at work, which led to the banning of asbestos in several European countries.

Certain laws and regulations apply to all economic activities taking place in a given territory, whereas others apply only to certain sectors. The regulation may become so heavy and restrictive that companies see it as a serious hindrance to their activities, which, in a globalized economy, may encourage them to settle in other countries. Furthermore, regulation can be a source of opportunities and benefits for the company. Include protectionist laws of some countries for a long time limited foreign competition in the clothing industry or in the wood. Despite the opening of markets, there are still laws and regulations that severely limit the entry of new players in some sectors of activity: production quotas in the dairy industry, mandatory membership of the order of pharmacists to own a pharmacy locally maximum number of outstanding taxi permits in a given territory, allowed the Canadian Radio-television and Telecommunications Commission (CRTC) to create a new radio or TV station, and so on. Some of these protections are disappearing in many countries, but in the first decade of this century, they are still present because of the important role played by governments in the management of local economies. Maximum number of outstanding taxi permits in a given territory allowed the CRTC to create a new radio or TV station, and so on.

Also, the rate and nature of taxation (established by various government authorities) vary greatly from country to country, and it is always a matter of concern for those developing the company's strategy. On one side, there are companies that are little concerned with the socioeconomic development of the societies in which they have particular activities and looking to reduce their tax burden. To achieve this, they use various means even up to install their headquarters in tax havens like Liechtenstein or Nassau. This, however, is not the case for the vast majority of businesses, large and small: they keep their headquarters outside tax havens, behave as good corporate citizens and pay their taxes. However, before increasing their investment in a given country or to settle in a new country, business leaders assess the rate and type of taxation that exist in the country as well as tax incentives to businesses. If they decide to continue their activities in this country, they will analyze the impact of taxation on their activities and adjust their behavior accordingly.

A country that chooses to tax significantly payroll companies can encourage the latter to reduce the number of employees by resorting massively to automation, outsourcing or self-employment. The differences between countries are so important for business investment that some countries have begun to significantly change their tax systems to make it attractive for companies tempted by US states. Business leaders are aware of the importance of this political environment that we just described. That is why they seek to influence through various means: a contribution to the election fund of political parties supported activities of lobbying, bribes wine.

Otherwise, there is what might be called a true "market influence" with public authorities. Already, Dahl had a pluralistic conception of life of Western societies, as opposed to elitist conceptions that prevailed: in any society, several groups of power, and none of absolute power. The existence of counterbalancing powers is what Dahl prevents the domination of one group over others and allows the exercise of democracy. To act strategically, companies must have a good knowledge of the various stakeholders (stakeholders) that may affect the conduct of their activities. However, the analysis of the sociopolitical context is complex. First, it must take account of three levels—the societal level.

According to this author, the political context has three interrelated elements that managers need to consider in their analysis: (1) the structure, that is to say, the analysis of stakeholders, including their relationships, interests, their expectations and their capacity for individual and collective action, to understand the networks in which they are part; (2) dynamic, that is to say, the study of the evolution of the major social challenges lobbyists to associate; (3) logical, that is to say, understanding the reasoning of the closest partners to assess the possibilities for joint action with them.

Such analysis leads to the construction of scenarios to guide the interaction between the company and its context. In the current environment, companies are placing more and more importance to the image they project in society in general and to the authorities in particular. It is not an image that the company can easily build and manipulate using sophisticated techniques of image makers and communicators of all kinds, despite the belief of some leaders. This is basically for the strategic management of corporate identity that focuses on the company's relationship with its environment and particularly with its key stakeholders.

The strategic management of corporate identity is now part of the leadership roles, and companies that give special importance to this aspect of management, often put up a structural entity in which specifically deals.

If a company like Nestle had so many problems with the marketing of its infant formula in developing countries is that it did not address the controversy surrounding their dissemination as an attack on their identity or therefore, as a major strategic threat. The companies belong to industries characterized by the use of different technologies. A company that has the latest technology often has a significant competitive advantage. Japanese companies after World War II have understood. Aware of the technological gap, but determined quickly become important players on the world stage, they bought, copied and pirated technologies developed in Western companies. However, the competitive advantage of flowing specialized technology tends to fade over time, since all the companies are gradually learning.

To regulate competition "technological," countries have established the patent system, which allows developers of new technologies to be protected for a specified period. This is the case in the pharmaceutical industry. But the debate between original brands and generic products shows how the patent system is not considered sufficient protection by innovative companies that spend a significant percentage of their turnover on research and development new products. The technological environment is changing very rapidly. The advent of computers has upset not only production systems but also the methods of supply, distribution, and marketing of products and services, and research techniques. To track changes in the technological environment, companies often endow a system of "technology watch" whose data and analysis are critical when strategic choices have to be made. Companies are very sensitive to the state of the general economy. In developing their strategies, they take into account several factors, such as interest rates, exchange rates, inflation, and unemployment rates.

Central banks use interest rates to speed up or slow down the economy, and this has important consequences for companies. Thus, in a situation of low interest rates, businesses have access to capital at lower cost, while demand for their products increases because the savings incentive is low. It is therefore a favorable environment for the expansion and business growth. Exchange rates also have influence on business. For example, when the Canadian dollar is weak against the US dollar, it promotes export-oriented companies. Conversely, when the Canadian dollar is strong against the US dollar, imported products seem cheaper and it increases competition for local products.

In the current context, where there is a great connection between the companies, it is becoming increasingly relevant to talk about the operating environment as a business network. Interest in corporate networks began in the year 1960. This is when we started to be interested in a particular way to network and interorganizational relationships. According to Evan (1966), organization (focal organization) belongs to a network of organizations (organizational set). The knowledge of this network allows us to understand the autonomy of the organization's decisions, the forces that incite to compete with other network organizations to cooperate with them, and its ability to achieve the goals is fixed.

For Astley and Fombrun (1983), when network organizations are strongly connected to each other, they form a turbulent environment whose properties are independent of the action of each of the network organizations. In this context, the organization's plan to work together to absorb the variations present in the interorganizational context. The collective strategy is the result of collaboration between organizations of the same network. Frery spoke about networks of companies using the term "transaction structure": as opposed to a financially integrated structure, a transactional structure is defined as a composite organization, bringing in the same value chain, capitalistically autonomous stakeholders, linked by a series of recurring transactions.

We find networks of enterprises in all industry sectors. However, some industries are characterized by strong ties between their various stakeholders. In this connection, it often gives the example of the automotive industry and the linkages between manufacturers' cars and their subcontractors. Worth mentioning are Bombardier, in the aviation industry, and Benetton, in the garment industry. In the case of business networks and transactional structures, there is plurality of financially independent companies linked by a multifaceted and complex trading system. Each of these companies cannot be considered as an autonomous entity, with clearly identifiable boundaries, and having its own decision-making center.

The first question is: What type of link must unite firms in a network? The second question, more strategic, must also be raised: What is the role of different network companies in the formulation of the strategy? Then, in what circumstances does it develop individual strategies and collective strategy? Can there be real planning guidance to a network? How the network adjusts as a result of changes in the operating environment? Are there transposition into corporate networks, the top-down approach (literally, from the top down) that characterizes the strategic decision-making in many companies? What is the role of the different network companies in the formulation of the strategy?

Strategic management of a company belonging to a network character-ized by high connectivity requires leaders to rethink the traditional way in which they managed to establish their business strategy. They can make it by taking into account the strategy of other Entre taken with which the company works in a given industry. These can then be considered as a part of context.

CHAPTER 7

Advanced Operational Business Analysis

The business context is the context in which the actions of the organization are embedded. The context of the case is both a reality independent of the organization and construction of its leaders. Different organizations can see in the same context different dynamics, different opportunities, and threats. This means that there exists between the organization and the context of business, one relationship. The organization's resources then have meaning when placed in the context of the case that it "has chosen." Competitive advantage is defined and therefore built in reference to what is happening or what might happen in the context of business. Take the flower industry.

In the past, the industry included several intermediaries, including florists, pharmacies, and supermarkets so that consumers pay more than 800% of the price paid to the producer. The company (Calyx & Corolla) created a network so that the flowers can be sent to the consumer, cooler and at a lower cost. It has established close relationships with producers, helping them to find the best packaging materials and informing them of the status of stocks and demand. It also concluded an alliance with FedEx to facilitate the delivery and enable consumers to receive their product within two days after picking. "Calyx & Corolla" has become a central player in this industry 10 billion. Many observers believe that the use of e-commerce will accelerate this process of eliminating intermediaries, not only for flowers but for many industries, including fresh agricultural products. The opposite can also happen, that is to say, that competitive advantage can come from offering an intermediary service where the customer is subserved by the existing system. In the airline industry, which is relatively opaque to the consumer, many agencies specialize to provide intermediation services, especially to find the best routes and the best prices.

But the most spectacular example is Google. This company realized before anyone else the problems that the explosion of information through the Internet posed to users. She then developed a series of tools or search engines that can quickly find the information available on the Net. The general engine you do remarkable achievement that facilitates the search for any information publicly trace in any of the major languages used in the world and this in record time. In addition, Google has stepped specialty engines that allow, in a particular area, such as maps or languages, to find instruments that put a lot of time to locate on the network. The popularity of Google is able to simplify it has introduced. Apart from Google, many other engines and media have begun to respond to the needs of increasingly accurate. YouTube, for example, that allows publishing filmed documents, is one of the latest blockbusters.

In all cases, a sustainable competitive advantage is often a specific building, systematic, that takes time. This construction is influenced by learning that the organization has made in its history but it can also be modified, adjusted and redone to ultimately position the company favorably compared with its competitors. The American adventure of the Spanish company Terra Networks is revealing in this regard. Terra, a subsidiary of Telefónica giant, is one of the largest content providers and Internet services in the Hispanic world. Entre decision moved to the United States and is made visible by creating www.terra.com site. She quickly realized that the task was not easy, before competitors like Yahoo and StarMedia, or international competitors like El-Sitio and Loquesea. The problem is related to the fact that it must satisfy a complicated market, namely US Hispanics. The latter, although bilingual, are nevertheless also different from Americans as are the inhabitants of the countries of Latin America and Central America. Despite the small number of Hispanics on the Internet, Terra defends remarkably well.

First, to encourage use of the Internet, Terra has teamed up with the telecommunications company IDT, New Jersey, who was already a customer of Latin origin. In addition, Terra has become known as the company that best understands the cultural and linguistic affinities of Hispanics in the Americas. Not only Terra said, but she practiced. By visiting the US sports Terra could fall on a headline about Tiger Woods. But if you clicked on the link Peruvian, you could get information on the soccer club "U." In the pages of competitors, we would probably find the same title about Tiger Woods, but nothing on the Peruvian soccer club. Managing such diversity requires considerable work. Indeed, some Hispanics prefer to browse in English and others in Spanish. This forced Terra to provide local content and trade electronically in both English and Spanish. For this, it has managed quality alliances with the Miami Herald Latin America and MTV Networks and did the same thing in California, New York, Miami, and throughout Latin America. In addition, Terra produces original content as its site that allows immigration to interact with people who have recently emigrated or who know the immigration laws. Wall Street does not make a mistake. Terra shares have risen dramatically. In February 2000, they were worth 850% of the price of the initial public offering. The value continued to increase despite the upheavals and the crash of high technology.

These examples demonstrate that the development of competitive advantage requires first knowing who you are and know the context of business. Then it requires the construction patient, systematic and specific resources and competencies that differentiate the company from competitors. In this chapter, we will reveal what lies behind the incredible creativity of companies and present useful methods of analysis of resources and internal capabilities. In the first section, we will propose traditional marches, to lead us in the second section, to newer methods of value analysis, including the value chain and the organization's conceptualization of ideas as an assembling resource, capacity, and skills. In the third section, we discuss new and promising ideas for creating patterns of recognition and value retention. Traditional analytical approaches based on the idea that the context of business is easily recognizable and understandable. Just by identifying key success factors, that is to say, what to do to be successful in a particular industry, to determine the gap between resources and capabilities and what is required and, finally, try to reduce this gap. One of the first approaches used to examine the business of discovering and specify the forces (what we do better than its competitors) and weaknesses (so underperformed its competitors) of the organization, by connecting them with external analysis,

This approach, popularized in the "Strengths, Weaknesses, Opportunities, Threats," is widely used by practitioners of strategic analysis, who find both easy to understand and convenient to use. Force is a resource or an activity that an organization is particularly good, better than its competitors. This is a feature that gives the company a special ability. It can be a special skill, expertise, a resource available to the company exclusively, or a reputation that the company has built over the years. It is obvious that a force can be tangible, such as availability of funds, or intangible, such as name, reputation, and technological know-how or managerial, including the ability to innovate and stand fast in the market with new products. Thus, when talking about Alcan, it refers to its ability in identifying the sources of raw materials, government relations as well as production and marketing, a combination of capabilities difficult to imitate. This is also a barrier for any new entrant. 3M has many distinguished itself by institutionalizing innovation and by making the source of real advantages over its competitors. Sony, meanwhile, has made its ability to create products "an insurmountable barrier" for most of its competitors. It refers to its ability in identifying the sources of raw materials, government relations as well as production and marketing, a combination of capabilities difficult to imitate. This is also a barrier for any new entrant.

A weakness is a lack of resources or performance of critical activities is lower than that of the competition, making it vulnerable between taken by its competitors. It is important to be aware of his weaknesses to guide strategic choices and avoid strategic ways in which the business would be less strong than its competitors. To identify the strengths and weaknesses of the organization, one can make a diagnostic analysis by external people, whose mission is to examine critically and comparatively the organization's practices. One can also conduct a survey of key managers and discuss situations for which there is disagreement. To guide the process of identifying strengths and weaknesses, Stevenson (1976) provides a review canvas large business function. This framework assesses the company's position in relation to each of the elements, from the perspective of managers or from the perspective of former dull experts. We can then examine the situation of the company, comparing it with that of its competitors, as perceived by managers themselves or by external experts. For example, we often asked managers to note the company or its competitors, each of the elements of the canvas, using a scale of 7 or 10 points. Each element is then distinguished from the others by a weighting system that reflects their relative importance to the company. Normally, the total gives a good idea of the strengths and weaknesses of the company. However, we must use this table with caution. Indeed, identify strengths and weaknesses involves both the review of overall scores, but also the examination of the scores along each line.

The pattern of strengths and weaknesses is the basis of the portfolio analysis of traditional products developed. Indeed, these analyses are all based on a representative dimension of the external, such as market growth, and a representative size of the house, such as relative market share. The company's success, therefore, requires a unique combination of characteristics of the business context and that of the company or the analyzed strategic unit. The pattern of strengths and weaknesses is also the basis of the PIMS (profit impact of marketing). PIMS is born of General Electric executives' desire to better understand the advantages and competitive disadvantages of their strategic business centers (business units).

This benchmarking model is valuable when trying to assess the strengths and weaknesses of an organization. Using the model of the strengths and weaknesses can still be sharper if combined analyzes such as the experience curve, lifecycle or growth vector, we look further. These analyses are otherwise also useful in complex situations. The best known are those generic strategies of differentiation and cost leadership.

Strategy on the costs normally helps to cope with the competition because it is to reduce prices and to gain greater market share. In return, a larger market share can produce larger quantities and, therefore, can reduce costs below what competitors are doing that have a lower market share. The principal of this cost reduction based on volume is described in the model of the experimental curve. This is one of the most popular models in strategic business management. Because of economies of scale, directly related to higher manufacturing volumes required to meet the market demand, and also because of the experience gained during the production process, the costs will be lower for companies that have the highest market share. Although cause and effect are not so mechanical that might suggest there is a relationship between the size of the facilities and production costs.

The economies of scale from the fact that, when increasing the size of facilities, investment and operational costs do not grow proportionally. Thus, an oil refinery with a capacity double that of a competitor can only have 20% more in fixed additional total costs of investment and operation. In ramming the unit cost, it is clear that more facilities are large, more the cost per unit decreases. Beyond the savings in production, there may also be savings in commercial expenses, research, and development and administration. The use of a learning curve as a basis for choosing a strategy focuses on costs. This strategy requires standardization of production processes and specialization of labor and manufacturing tool. This increases efficiency, certainly, but reduces flexibility.

Some important changes to the manufacturing process, due to innovation, can even put the company in danger. The concept of the experience curve should be used with caution. So, try first to increase production capacity to achieve significant market share can exert considerable stress on the operation of the business. For example, from 10–30% market share in a market that is growing 15% requires growth of 43% for 5 years! After 5 years, the production volume has increased six-fold! We must also understand what applies to the phenomenon of experience. In the automotive industry or the aviation industry, the final product is a combination of hundreds of components, each having a different experience curve. To make informed decisions, it is very important to understand how these experiences combine. Finally, it is sometimes useful to think of a shared experience, when an alliance can afford to take into account the production volumes of two or more companies at once. This is the case when companies like Renault and Volvo team up to build common engines. The product life cycle is concerned with the normal development of the company's products in their market. A product is first launched and the result is a gestation period before it is accepted by the market. If necessary, the request first experiencing strong growth, and growth slows to enter a mature phase,

In general, as all products familiar with this demand pattern in the market, we can organize the activities of the company accordingly. It does not support the activities in the same way as the product is in a period of growth or maturity. When the product is in the introduction phase, the technical aspects, such as product development, dominate. When the product is in the growth phase, production issues take over. When growth slows, the activities of marketing and distribution necessary to maintain or enhance market share and increase margins. Finally, the product decline, we must "harvest" and reap the profits. The life cycle induces specific business strategies for product portfolio balance.

At this stage, however, we must emphasize that the life cycle is a given on the natural behavior of all products, but that the details of the life cycle of a product, as the length of each of its phases, depends a lot of the actions of companies in the industry. Thus, it is expected that the life cycle is longer in an industry where companies prefer product improvement rather than changes in products and focus on the control of market domination by costs and barriers to entry, advertising, and promotion. Conversely, in an industry where companies are driven by innovation and tried to change their products regularly, the life cycle can be very short. We will succes-sively address two widely used methods in strategy, namely the model of the value chain and the approach of resources and skills. In his model of competitive analysis, Porter suggests examining the company using the concept of a value chain. The value chain can be defined as the set of distinct activities contributing to the creation of value that the customer is willing to pay. Going further, it is to examine the sequence of activities of a company in order to understand how they are used (or could be used) to do business differently or better than other industrial companies.

Porter suggests examining the company using the concept of value chain. The value chain can be defined as the set of distinct activities contributing to the creation of value that the customer is willing to pay. Going further, it is to examine the sequence of activities of a company in order to understand how they are used (or could be used) to do busi-ness differently or better than other industrial companies. Porter suggests examining the company using the concept of a value chain. The value chain can be defined as the set of distinct activities contributing to the creation of value that the customer is willing to pay. Going further, it is to examine the sequence of activities of a company in order to understand how they are used (or could be used) to do business differently or better than other industrial companies.

In this model, it comes to determining what business activities create value. To do this, we decompose operations into simple elements to better understand how each of them contribute to create value to customers. The model distinguishes between primary activities, such as production, marketing, logistics and delivery, service, and support activities, such as procurement, technology development, human resource management, and infrastructure (general management and associated services). For Porter, the most important competitive advantages derived from the differentiation and the ability to have low costs. Examination of the value chain to better understand how each activity influences differentiation and cost. This is called determining the drivers of differentiation or cost. Porter suggests that it is in the arrangement of activities that the company finds original ways, sometimes difficult to copy, to stand out from the competition and build a decisive competitive advantage.

This is what distinctive competence of the organization is. A synthetic view of the value chain must be presented. If we talk about "value chain" is that activities are related and form a coherent whole. Moreover, the components of this set must be maintained in connection with the coordination activities. Note that coordination can itself be a source of competitive advantage. It is for this reason that system management is an essential element of the value chain. Once the elements of the value chain are completed, it is possible to assign a cost to each of the elements to better appreciate their contribution to the total cost of the finished products. We can also consider the contribution of these elements in terms of differentiation. Knowledge of the contribution to the cost or differentiation, or both, compares the company with its competitors and thus to seize the company's ability to support its strategy. Take the example of the costs leadership strategy: To achieve this strategy, the company must be able to be produced at a lower cost than its competitors, or to procure the raw material for the best price, is to ensure deliveries at very attractive prices, or take advantage of other benefits in terms of costs, or to have a combination of several of these activities.

Let us return to the article where Prahalad and Hamel (1990) suggested that successful companies develop sustainable competitive advantages tend to see themselves as a portfolio of skills rather than as a portfolio of business centers. In 1980, GTE posted sales of 10 billion, while NEC, much smaller, realized a turnover of 3.8 billion. The two companies had similar technology and computer basics but GTE exercised in addition to telecommunications activities, particularly in the field of telephony. In 1988, GTE reaped revenue of 16.5 billion, and NEC, to 21.9 billion. GTE had become a phone company, keeping only a few activities in the areas of lighting and defense. It had disinvested by selling Sylvania, its activity in the field of television and Telenet. It was also involved in joint ventures for its switching operations (switching) of transmission and digital PBX and had left its semiconductor-related activities. Therefore, its international position had deteriorated from 20–15% of its turnover.

By comparison, NEC became the world leader in semiconductors and one of the most prominent for telecommunications products and computer players. In addition to the mainframe, it produced mobile phones, fax machines, laptops, making the link between telecommunications and office automation. Among the five largest telecommunications companies, NEC was the only one to also work in the semiconductor and mainframes. NEC has established an explicit strategy to exploit the convergence of computers and telecommunications (called C & C). The company believed that success would come from the acquisition of specific skills, particularly in the field of semiconductors. The strategic architecture that resulted was then circulated widely within the company and outside. C & C committee was formed to oversee the development of skills and core products. The company has devoted significant resources to the consolidation of its position in components and central processors.

The study by NEC led her to think that computers would evolve central processors to the decentralized process. She is convinced that gradually the activities related to components, communications, and computer technology would overlap so that it would be difficult to distinguish. A company with expertise to serve these three markets would, therefore, benefit difficult to imitate. This has, therefore, led NEC to invest heavily in semiconductors and to form a multitude of alliances for the rest of its activities, including Honeywell and Bull, in order to "avoid developing what already exists." Meanwhile, GTE continued to regard its activities as autonomous entities, regardless of skills that would enable the company to better position itself in future markets. This attention to the skills and central resources, as well as central products, (those which control is essential to succeed in the relevant markets), is clearly different from the traditional tendency to consider the company as portfolio products or activities (SBU or CAS) relatively autonomous. In fact, many companies are trapped by the dogmas of the autonomous management of CAS and by the pressure of short-term performance.

Although performance and short-term competitiveness are dependent attributes the price-performance ratio, the global competition imposes standards ever higher in terms of cost and quality.

Companies who benefit are those that are able to build the skills, at a lower and faster, than their competitors, cost. The real source of competitive advantage is the ability of managers to manage the process by which technology and know-how are transformed into skills that enable adaptation and seizing opportunities inaccessible to others. In sum, if the products are leaves, flowers, and tree fruits, skills are the roots. The more one observes the process of consolidation and development of skills, the more we see the importance of organization and coordination. Sony (Kettani, 1996) emphasizes the importance of its technologists, its engineers and traders share the same understanding of customer needs and technological possibilities. This allows all the key elements of the organization to keep pace. The central competence is so communication, involvement, and commitment to work without worrying about traditional organizational boundaries. This is what allows new businesses to expand globally.

What's interesting, say Prahalad and Hamel (1990), is that the core competencies do not diminish with use. Instead, use and sharing make them grow. However, we must protect them, or they can disappear, if not applied in the right place. One can also leave from the critical skills for the future. So, it is likely that GE, although this reflected a deliberate strategy, not only sold to Thomson of electronic activities for the general public (in the field of television in particular) but also the essential skills of a sector likely to experience strong growth.

To determine the core competencies in a company, three questions must be asked:

- This skill has access to a variety of markets? Thus, competence in display systems allows participation in markets as diverse as calculators, miniature televisions, monitors, and laptops, auto dashboards, etc. Just see the strategic behavior of Casio to be convinced.
- This skill does contribute significantly to the creation of value for the customer with respect to the final product? Honda's expertise in engines or that of 3M in specialty adhesives used to positively answer the question.

Is this skill difficult to imitate by competitors?

It is conceivable that a company can really develop global value skills in five or six major areas. Although we may list a lot of skills, we must try to condense into blocks. This is also what allows to discover the missing blocks and those that allow for alliances. Prahalad and Hamel have stimulated a generation of researchers who wanted to know whether there was a systematic relationship between the nature of the skills and success. This gave rise to what is called "resource perspective." This work was launched by Wernerfelt in 1984. According to this perspective, competitive advantage comes from internal resources, and it is better to focus on resources rather than the dynamics of the industry, considered too volatile if we want to stand out from competitors and provide sustainable benefits. Resources are thus broader expression skills. Although this term includes everything that can be used to sit a competitive advantage, it includes both material and tangible resources that intangible resources such as skills.

These have a more systemic nature and the result of "the interaction between technology, collective learning, and organizational processes." It includes both material and tangible resources that intangible resources such as skills.

Thus, Sony's capacity to generate innovation or those of GE facilitate adaptation to change and combine skills are a number of tangible and intangible resources. To achieve sustainable benefits, resources must possess a number of characteristics: It is necessary that a resource is rare, durable, as the company is appropriate and it is difficult to imitate or substitute. The skills generally all these qualities. As they are related to processes that take a long time to produce results, they are of considerable importance. When the source of competitive advantage, it is hard to beat because, for the same skills, competitors must follow similar processes and take the time necessary to achieve it. The commitment of the employees, the reflexes they have in their interactions, coordination, and integration facilities, the ability to adapt and the ability to innovate are all skills that are slow to build but are difficult to conquer fortresses.

Miller and Shamsie (1996) developed some resource perspective with links between the nature of the resources, the nature of the business context and business performance. They, first, proposed to distinguish resources "based on the property" and "knowledge-based" resources. The first encompasses all resources protected by laws or regulations. Thus, a patent owned by a mine or a property can provide a competitive advantage to the (individual or organization) who is the legal owner. As for resources based on knowledge, they approach the definition that we give skills. These are skills that are not available to others. It is normal to think that resource properties are most effective when the context of business is stable, whereas resources-knowledge is especially effective when the business context is turbulent. Miller and Shamsie have especially demonstrated in the case of the strategies of Hollywood studios.

The resource theory is in full development and to test systematically the relationships previously little understood. In a study in which we participated, we linked the nature of resources and the nature of the context of dealing with the existence and performance of cooperation strategies: Stable and homogeneous industries with little uncertainty tend not to generate cooperative arrangements unless the dominant games and resources based on the property are compatible and complementary. In such cases, cooperative arrangements generate superior performance. When cooperation is attempted, despite identical or incompatible resources, it leads to poor performance. Unstable and heterogeneous industries with a high level of uncertainty, generates a large number of cooperative arrangements unless the peripheral games and resources based on knowledge are perceived as incompatible or identical. In the first case, the cooperative arrangements generate superior performance. In the second case, the cooperative arrangements that are tempted lead to poor performance.

Resource theory draws attention to the importance of internal resources when you want to generate a competitive advantage. Even if the resources cannot be considered without reference to the context of the case, the theory suggests that we really have to leverage our resources and it is on this that the focus should be. This perspective leads us to revise the strategic analysis, putting the center what we have and what we know to do, and by modulating the use of resources and skills to meet the changing market needs and requirements of the context of the case. A supportive manager to use this approach find valuable approaches proposed by Prahalad and Hamel (1990), most of which were mentioned earlier.

As we have repeatedly said, the strategy is an art. The complexity of the business is so large that, firstly, it is difficult to understand what is happening and, on the other hand, there is room for many successful strategies. Many researchers and consultants have sought more or less awkwardly to discover strategic responses to different situations in the context of business. This goes against the idea of creating art involved. It remains that the examination of practices that lead to success is a relevant exercise for the manager, as it is relevant for aspiring tennis players, for example, to study the performance of great players. It is therefore proposed to look at a few examples of success. There are theoretically an infinite number of behaviors that generate profit.

We present 5 of them, which are related to different aspects of the value chain: (1) changing the value chain; (2) increased attention to customers; (3) modification of the distribution channels; (4) the original management of products or services; (5) knowledge management. You can change the value chain in different ways: Value chain can be disintegrated or reinstated; it can be compressed, with a decrease in the importance of one or the other traditional activities to create value, or extent, with the strengthening of a link, was weak.

Nike, Benetton, and the Chinese company Li & Fung were the champions of disintegration. These companies have quickly realized that in the value chain, some features proved crucial, whereas others played the role amenities. Li & Fung, from a small textile company, has now become a diversified supplier, were able, after long years of computerization efforts, to focus on negotiating with the end customer, product design, and management of the supply chain, leaving to others the essential shareholders of manufacturing. In doing so, the company has dramatically improved the length of the procurement cycle clothing and electronic equipment, thus gaining a competitive advantage difficult to imitate. The disintegration may also leave room for shareholders working in a specific sector.

Thus, in the telecommunications industry in the United States, Qwest Communications has decided to be, "the carrier's carrier." By building a national network of optical fiber, it resells its services to local suppliers as GTE and US West, which then sell them to the public. 10 years after its creation, Qwest was worth 8 billion dollars. The reintegration can also be a source of competitive advantage. The pharmaceutical industry is often cited as an example. The change of power relations in favor of distributors sparked a consolidation process, initiated by Merck.

This company has acquired Medco, a company managing pharmacy benefits, serving employers and large buyers such as hospitals. Many other companies in the industry have followed suit. Gap, the largest sportswear retailer, has evolved from a jeans shop in a chain of 2000 stores serving four major markets: Gap (midrange), Gap Kids and Baby Gap (children), BRepublic (average and upscale), and Old Navy (midrange). Gap is now the design of its own products and maintains a stronger connection with customers. These companies have all managed to grow better than their competitors, in terms of both sales and profits. Sometimes, success can come from the extension of the value chain by improving capacity "neighbors" that prevent the company to create value. Problems are often cited the fact that McDonald's had in his rea- fragmentation, and lack of storage standards in the potato industry in Europe. The company has taken years to consolidate suppliers, train them, provide them assistance and standardize systems.

This expertise served him later in many emerging countries in transition, such as Russia. Toyota and Wal-Mart have done the same thing in many sectors. In particular, Toyota built its relationship admirably system with suppliers and created significant competitive advantages, such as those associated with "lean" to complete the "just-in-time." In an industry, the value can migrate upstream or downstream according to circumstances and strategic behavior of companies. The concept of the center of gravity (Galbraith, 1983) has been proposed to describe the ability of a business to adapt to these changes. In the 1980s, IBM was the very ring in PCs, the lower ring being Intel. IBM had also decided to invest in Intel (19%) to help strengthen. Later, the reverse happened, IBM being the weakest link and Intel's strong ring. This example demonstrates the value of migration along the value creation chain in an industry.

Customers are the ultimate arbiters in terms of value. Their behaviors can create or destroy. So, there are situations where the microsegmentation or redefining the customer is required and must be aware that customers with power relationships are constantly evolving and changing opportunities for value creation. To better meet customer needs, companies can make the microsegmentation and attempt to meet the specific needs of each client. This was a utopia in the last few years, but thanks to new information technologies and communications, we get to do more. So, Levi Strauss Company discovered that many of his female clients were upset to have to try 15–20 jeans before finding the right size. The company was developing, in 1994, the Personal Pair program, which allows a client to quickly identify models that suit them. In 1997, the program generated 25% of sales in Levi's stores.

The following year, Personal Pair was replaced by Original Spin, holding accounts also men. This has provided a better understanding to customers; Levi-Strauss can now offer in the stores over 750 choices of adjustment. The following year, Personal Pair was replaced by Original Spin, holding accounts also men.

Microsegmentation is stimulated by the diversity and sophistication of larger and larger customers, demanding more personalization and more choice. It requires technological development that serves different segments effectively. Another approach is to redefine the customer that wants to serve. This is true of Bang & Olufsen, a European company of electronic products, including traditional clientele consisted of audiophile connoisseurs who appreciate the technological sophistication and product design, but that does not generate profit. The company had to redefine its customer base to include those seeking luxury and are sensitive to elegance and status. She began selling its products in emphasizing the exclusive character of the B & O products, gradually managing to straighten a threatening financial situation.

From 1989 to 1997, the ratio of the value of shares on sales rose from 0.2 to 1.5, whereas the ratio was around 0.5 for most consumer electronics companies. This is the traditional methods of pressure strategy (marketing push) and pull strategy (pull marketing), but many are rediscovering them with happiness. Thus, in the 1980s and 1990s, DuPont, to the reluctance of its immediate customers toward his master Stain-product, has practiced a strategy of attraction (pull) by creating demand among consumers. Intel has done the same for its microprocessors. Gradually managing to straighten a threatening financial situation.

Changing distribution channels can be done in different ways. We can increase the number of specialty channels or, instead, focus on a few channels to achieve economies of scale. It can also reduce the number of steps in the distribution or interpose an intermediate where there were none before. The case of coffee dispensing illustrates the multiplication and specialization of channels.

Today, coffee is sold in many places with chains increasingly special-ized offering opportunities to drink or buy differently coffee. In the food distribution sector, we have witnessed the opposite phenomenon. It started by opening hypermarkets which destroyed much of the fragmented system of small traders. The phenomenon has spread around the world and took surprising forms. Thus, the concentration did not hit small convenience stores, but those who supply them, leaving the convenience store to better serve customers in the area. But even for convenience, the Couche-Tard chain is changing the landscape with its dominance of the convenience store in North America. This concentration phenomenon affects many industries. Thus, creating retail chains, Blockbuster (in the field of DVD rental counters) and Barnes & Noble (in the libraries sector) have managed to offer a better selection of the most appropriate time, more attentive service and sometimes a better price. The distribution channel can also be reduced or disappear, as was demonstrated at the beginning of this chapter by the example of Calyx & Corolla in the flower industry.

Conversely, a new intermediation mechanism may occur when the client does not all the information it needs or finds it difficult to pick itself the product or service desired. So, the formation of Creative Artists Agency has allowed Hollywood studios to get the stars of services, writers, and directors, without being obliged to do so in the room. The artists have also gained something since their bargaining power has increased. The whole system has gained stability and predictability in this new way.

Many companies focus on trademark or try to focus on promising products (blockbuster). They create situations where products can become the basis of development and increase profits. They can also create a kind of hierarchy of products in order to seize all possible profits. Finally, they change the product idea to the idea of service or solution. When the customer has too many options on the table, he is frustrated. The differentiation by brand overcomes this problem. The brand becomes a sort of guarantee of quality, reliability, etc. So, if we can, like Coca-Cola, Swatch, Evian, or Intel, make a mark, we can reap the benefits long. In the early 1990s, two identical cars were built at Nummi, a company born from the merger of Toyota and GM based in Fremont, California. Although both cars were identical, the one who wore the Toyota brand was selling faster and $400 more on average than the other car.

In many industries, the profit migrates a product portfolio to some carrier's products (blockbuster). We must recognize and create better products. This is done in a very large number of industries: Film production, pharmaceuticals, music, books, real estate, sports talents, television, etc. An excellent model for the development of promising products comes from the pharmaceutical industry. Since 1970, Merck has developed this system and introduced in 1981 Vasotec (a drug against hypertension), among other drugs. Today, every company has its carrier product: Schering-Plow and Claritin, Eli Lilly and Prozac, Pfizer and Viagra. Disney, Bloomberg, Sony, Michael Jordan or Tom Peters have in common is profit multipliers. When Disney created The Lion King, she followed toys, clothes, books, TV shows, music selections, and ice shows, significantly increasing its profits. Similarly, Michael Jordan used his personality and qualities of basketball and Tom Peters has used his talent motivator with business people to significantly increase profits by a range of other activities.

American Express has done much the same thing but by structuring deals in a pyramid. So, she put on the market a Gold card with an annual membership fee of $100 and, later, a Platinum card with an annual membership fee is $300. She did the same for his other services. At Gillette, we can recognize the same desire to offer a tiered range of shaving products. We talk about fewer and fewer products and more solutions. Thus, in 1990, GE sold to British Airways a contract of the "engine." BA benefits from the use of motors which are the property of GE, which maintains and follows them throughout their useful life. GE is also transforming all its products into solutions." Similarly, Honeywell offered to Boeing successfully not to assemble subsystems designed by Boeing but take responsibility for the design and supply of all the avionics systems. However, it should keep in mind that solutions are never final and they are constantly evolving.

Knowledge is at the heart of competitive advantage, but it is not always used by the organization wisely. Some examples of strategic choices allow us to highlight the importance of knowledge and its proper use by the organization. The product is an invaluable source of information on the client. The retailer is often lost in the huge amount of information that customers generate behaviors, but the manufacturer of a product has the ability to track the product and its variants, to accumulate valuable information on customer behavior and thus achieving the following three objectives: (1) Effective management of product categories (called SKU, stock-keeping units, or inventory management units) in stores; (2) precision merchandising: Wal-Mart has developed a remarkable ability to understand consumer behavior, permit, as to avoid too high stocks or breaks. In addition, Coca-Cola is currently testing telemetry systems for monitoring the stock of each vending machine bringing precise management of these segments; (3) An increase in the success rate of innovation, with a detailed knowledge of consumer behavior. GE, for example, has developed sophisticated models of product performance in the context of use by the customer. This information is the source of innovation in products or services.

The activities in several sectors, such as hotels, bookstores, steel, aviation, etc., do not always generate the desired profits. However, they contain a lot of knowledge that can be systematized and sold with considerable margins. This gave birth to the management and sales companies' know-how rather than products. Thus, Marriott focuses on the provision of hotel management services. Barnes & Noble offer their services to manage the libraries in universities and other community locations. Similarly, the Japanese have always sold their know-how in steel making in Latin America, Korea, and elsewhere, when the industry began to decline. And American Airlines has operated its operational capabilities to develop a great competitive tool that was its Saber reservation system. The reverse is also true. One can spend fine knowledge on a topic or process to successful products. SAP (integrated management systems) and PeopleSoft (human resource management system) have created packages that have benefited from expertise developed through custom work done for clients. Knowledge was then converted into a product that allowed for the mainstreaming activities, but much more efficiently and at a lower cost. In general, we say that the major professional companies can grow by 15–23%, but companies like SAP or PeopleSoft.

So far, we have presented a number of tools to establish the internal analysis of the company: Analysis of strengths and weaknesses, identification of skills and resources, and the value chain. These tools are widely used by companies when formulating strategies and items that are of interest are often tangible elements or functions of the company. But there are other less tangible elements, such as leadership and culture, which are important because they orient strategic action.

CHAPTER 8

The Implementation of Business Optimization Mechanisms

It is presented here as part of the primary analysis, the one that used to define, design, and make (the terms are used interchangeably) business optimization mechanisms. The decision optimization is either an end or a conductive vein, a mediation mechanism with the context of business, a combination of internal resources to achieve a competitive advantage, the expression of leadership values and community people who constitute the company. In the analysis to design the applied management mechanisms, we find then inevitably these different facets and can integrate. To address the analytical framework must illustrate the optimization process through the real case of a now-defunct company.

This is food distribution of a large company, which had invested money in launching a chain of general merchandise stores. Years later, the company lost hundreds of millions of dollars trying to save the company. The history of this company appears to be an uninterrupted series of improvisations that led the company into a vortex so incoherent that nobody knew how to make sense of such an adventure. The company embarked on the adventure of general merchandise stores for different reasons. It had empty space she wanted to fill. She also had the confidence typical of a company that had known only success in its food business that could create a sense of power.

Thus, the same reasons that made the success of the company, including improvisation, also contributed to making inconsistent and dangerous decisions for the survival of the entire company. This suggests that improvisation, even great, is appropriate only when the degree of complexity of the organization is to measure the cognitive abilities of the leader. As soon as the company grows, improvisation must incorporate more systematic reflection and collaboration to the understanding of the evolution of the company and its business context. The lack of direction has been detrimental to the company. The new leaders of the company, have understood the need to do what should have been done earlier: Define the company and its rationale.

To start their strategic thinking, the new team needed information. It was first necessary to know the company by consulting clients. The question was to know why customers were loyal to the company. It was also determined competition: What rivals? The study of industry and market has entered a much-segmented market with competitors who tend to want to distinguish themselves from each other. Four major categories of stakeholders had emerged. General stores, large department stores seeking to attract a bit everybody, present in one place with quality products and different prices.

Then, discount stores, whose approach was based on the price or more precisely on the price-book. There were also "buying clubs," a kind of super-discount stores but the range could change at the discretion of procurement opportunities. Finally, there are companies that stood out by specializing. Specialty stores covered as segments where quality and status were, for the consumer, important factors to consider, that the segments where price and value were the first items to consider.

The customer profile a bit surprised: Young families with incomes above the average. Customers were relatively faithful and, more importantly, they saw the company, a store that was similar to department stores, regarding the assortment, fashion, and quality but with prices comparable with discount stores. However, the layout was less attractive and less good service in department stores. Finally, in terms of performance, the most problematic stores (in terms of deficits) were those distant. It also showed that the least requested products and less profitable were small appliances, whereas the most promising products were clothing. Using this information, it was necessary to define what the company is and what it should become. Knowing that it is easier to create a new company (we have more elbow) than flying an existing business (account must be taken of what exists).

It should be emphasized the importance of considering its own resources (the resources invested, debt, and availability) when defining what we want to do. In short, leaders gathered the information they had on competition, clients, and past results, and tried to see how they could "re-engineer" their business in the light of such information. They came to the conclusion that to optimize the company they should strive to meet the image that made them the most loyal customers. Also, it was necessary to improve the quality of the service, delete the store shelves, equipment that customers preferred to purchase elsewhere and concentrate activities on a regional scale. It was necessary to achieve this transformation format the staff, develop local, administrative reorganization, etc.

Finally, we had to change the company name. This exercise, very logical and very systematic background, has not produced the desired results, because, first, it was too late and because, second, the management team has not enough time to implement it. The disappearance of the headquarters, following a dispute between major shareholders, caused the premature liquidation of the company. No one was internal conflict to the board; the company would have been optimized and saved. This story shows the importance of defining the rationale and purpose of the business. There is no doubt that the absence of a clear definition of what the company has contributed to its liquidation. It is very important to adopt a systematic approach when deciding the purpose of an organization.

In applied management, there is a constant debate about the usefulness or the need for a systematic approach. Some stressed the risks of systematization, including rigidity and conformism that may result. However, note that once a company offers more than one product and is active on more than one market systematic thinking is to encourage convergence. A systematic approach to optimizing reflection often determines the productivity and business dynamics. Also, when the purpose is no doubt in the minds of employees in situations of complexity, the approach could be cons-productive. The analysis models of decision optimization suggest that the optimal decision, which is basically a rational form of decision-making to increase company performance, includes at least three steps: (1) An analysis of the business context to understand opportunities and threats it holds that he introduces; (2) an analysis of the skills and resources that the organization can use to take advantage of opportunities and deal with threats; (3) choice (after these two analyzes), among a number of options, those most likely to lead to better performance. Empirical analyzes involve two other factors: Leadership and social responsibility. It is suggested that the characteristics of leadership and values influence the choice of the latter and also the interest of companies should encourage its leaders to take into account the societal context and to make arrangements for the actions to be socially responsible.

This design optimization practices bring us he seems useful nuances to understanding the concept of management and optimization to improve its use by practitioners. The approach we propose allows integrating often-overlooked considerations. The optimization mechanism cannot make in light of the following four elements: (1) The context of business; (2) Resources and internal capabilities of the organization; (3) the desired social contribution; (4) value leaders. Here, the context determines the business opportunities and threats. The business context is a critical element in effect. This is, of course, the relevant context, one that is of paramount importance for the conduct of business operations (first includes major players with whom the organization is in competition). For a company, competitors make up the bulk of its concerns. But even a nonprofit organization is often in competition with other organizations.

What the "competitors" has a direct effect on the present and future health of the organization. The context of business includes all the shareholders, that influence the competitive game or whose mission is to influence this game industrial economy suggests that these shareholders are. Customers themselves, when they have an organized power; suppliers, when they have significant market power; and newcomers, who, due to low barriers to entry, can heighten competition. These shareholders maybe substitutes' manufacturers, who may have the same function as the products or services industry in a group of applications and determined that, in certain circumstances, they can behave as direct competitors.

The context of the case also includes governments, both as regulators, so as masters of the game, and as direct stakeholders (suppliers or customers). Thus, in the pharmaceutical industry, the government plays the role of guardian of the entrance. In the airline or telecommunications industry, it works to reduce the barriers to entry by eliminating rules that protect or enhance existing players. Moreover, it creates rules for the protection of citizens (corporate responsibility and professionals to the quality of products or services, for example) and the environment, as such, it requires all stakeholders. The context of the case is in constant motion. Like bees, multiple stakeholders continuously working to change the game to suit their own purposes. In doing so, they create situations that can be dangerous or threatening to the health of the organization and at the same time, they give rise to opportunities that can be exploited favorably by the company. The context also produced changes that are attributable not to a particular actor, but a large number of players, leading to changing historical, such as major changes in the economic cycle, major demographic changes, big changes sociopolitical, and cultural or major technological upheaval. They create situations that can be dangerous or threatening to the health of the organization and, at the same time, they give rise to opportunities that can be exploited favorably by the company.

Take the example of a manufacturing transportation equipment company, utility aircraft, small-range commercial aircraft, private planes, and corporate leisure facilities or snowmobiles and watercraft. In each of these areas, the company will face in the course of its existence, in various contexts. Thus, in terms of cars, buyers should be mostly local or national governments. In the field of air transport, it will live and will be negatively deregulation waves, as both aircraft manufacturer and as a subcontractor. As a manufacturer, it will be subject to safety regulations imposed by the country.

Internationally, the sociopolitical upheavals and the effects of political struggles should be a constant concern. The subcontractor for major aerospace players will bring the company to learn to live with powerful customers. In the area of rail transport, its government clients will also be particularly influential. In the leisure facilities (snowmobile and watercraft), it will have to suffer intense competition from new Japanese and American players: It will see the market decline dramatically when the oil crisis will make these more expensive equipment to operate. As a consequence, a separation of the sector and its consolidation into an independent company will be desirable.

Finally, it constantly lives with all kinds of substitutes for its products, which led him to focus consistently on the decision-making processes of its major customers. Understanding the business context then enables strategists and company analysts to determine, and sometimes anticipate, opportunities and threats as they arise and to appreciate their significance and evolution, to better optimize management practices. That said, associated with internal capabilities and resources available, these are the weapons available for competitive business optimization. Indeed, to do something lasting, it must consider the resources, skills, know-how that is available or can arrange to survive the pressures exerted by the context of business. The decision optimization can then move from analysis of the strengths and weaknesses of the organization. These strengths and weaknesses are of different natures:

- There is, first, the staff. It is the repository of the most important know-how, those who can differentiate the company from the competition. The expertise may be technical, administrative, organizational, and interpersonal:
  - Technical skills can be appreciated when analyzing the strengths and weaknesses of each of the specialties and functions involved directly in the generation of the value of marketed goods and services;
  - administrative expertise helps keep the company in balance internally, by the adjustment of personnel flow, equipment and funds, and with the environment, swinging efficiency and flexibility;
  - expertise interpersonal aligns the normal conflicts engendered by the action group; it facilitates adjustments and helps moderate the requirements of each and every group; it also facilitates the approach to the convergence of views necessary for decision-making;
  - organizational know-how keeps the rules of the game relevant and effective for the management of actions and initiatives that help the organization to function and adapt.
- The state of the technology and equipment is also an important capability, although this importance varies from one industry to another.
- There are also funds available or accessible.
- One can also mention the relationships and alliances with companies or influential groups of the context of business.
- Finally, practice and organizational functioning, with the structure, systems, and processes that dominate the life of the company and are difficult to change, are also critical capabilities. They can be positive or negative, depending on the competitive situation.

Each of these abilities can be seen as a strength or a weakness. This is the comparison with the situation of the main competitors that can say that. We can have skills quite remarkable and indispensable to be in business but, if they are accessible to all competitors, they no longer have much interest in the competitive struggle. By cons, if these skills are worse than those of competitors, it is in danger and you must either leave the area or work to improve them. However, if they are better than those of competitors, the company is in a favorable position and must be exploited by appropriate competitive positioning and by the expression of a challenging goal. From this point of view, here the example of a company producing metal cans for soft drink bottlers and breweries. It was the smallest of the four big companies in the sector. The two largest companies had almost four times the size. With limited financial resources, the leaders then worked out now so it is not really in direct competition with rivals. In particular, they exploited the company's strengths in the manufacture of filling machines and agility to respond to customers, by providing a set of services and products appropriate to resolve satisfactorily the problem of filling. In doing so, it no longer supplied cans, but met a critical customer need better than any of its competitors.

Therefore, it realized, the best performance on the market, far exceeding that of multinational companies and leaving far behind its competitors in the industry. Another example is a company that puts sanding discs that serve a particular market to polish and clean the metal surfaces. The products are standard and the market is dominated by three large international companies. These two merged to become the world leader. Economies of scale are considerable and competition is usually done by the price. At first glance, there is no place for a small business. Yet our company whose business volume is around hundreds of millions of dollars is a thriving business. To survive, it decided to approach the customer and examine how it used its sanding products. Based on a detailed understanding of the value chain of its customers, the products have been improved, with the help of a series of small innovative manufacturers internationally.

The company's salesmen have become advisors, able to help customers dramatically improve their use of products. In terms of value for the customer, the company provides products which, although more expensive units, are much more durable and, therefore, much cheaper to use. For those who follow product usage advice, the company guarantees a lower cost. This customer rapprochement course has a cost to the company since the company has more sales associates locally than its main competitors. In distribution, the company continued to use the existing network. Vendors have with distributors' close and fruitful books.

Dominated by large companies, they are happy to be heard and get even higher distribution margins. They, therefore, promote the company's products. Their advice to clients and go in the same direction as those of the company's vendors. They are happy to be heard and get even higher distribution margins.

Therefore, the company has gradually become the specialist in surface technology and, in addition to abrasives supplied cleaning fluids and cooling. Gradually, the company became the reference and the most respected player in the field of surface treatment. Thus, the company found itself in a position where it does not really provide a product, but a service that certainly included products (sanding, cleaning, and lubrication, in particular), but also a lot of important knowledge of the performance of work by the client. Customer proximity also allowed better develop products since the company was aware of the challenges and customer needs. Thus, in cleaning, Walter realized the cleaning of parts subjected to welding was done by the use of volatile solvents based on petroleum products, so that could ignite spontaneously and dangerous to the health of exposed personnel. He then strived to develop a biological product based on water and bacteria "eaters of petroleum products." This product, which has received many environmental awards, is trying to replace petroleum products and make the company a "softer" brand.

That said, in terms of optimization of the company, the social contribution is desired. The people who form the company may be concerned about the relationship between the company and society in general and its place there. The values and culture of the groups that constitute the company can result in an affirmation of the social role it should play. In the traditional language of the operational analysis, it is called "social corporate responsibility." Thus, a company can undertake several actions to protect the balance of the region in which it is active. Financing of social or sports facilities, cultural activities, and resorts opened to the public, paid release staff for local volunteering are very common. Other companies are also still occupied the socioeconomic and cultural balance of the localities where they have started their activities, including funding, are sports facilities, museums, and events of all kinds. Others have long financed the publication of newspapers and regularly financed the university and hospital, including several chairs in entrepreneurship.

It also happens that companies are sensitive to major social concerns and adjust their activities accordingly. Thus, history will record that some multinational companies had suspended their activities in South Africa to support the protest of the people against apartheid. Sometimes, well-established institutions undertake to discuss major social issues with the surrounding population. Some universities have responded to intense pressure from students and local people to divest from companies doing business in South Africa. They have long resisted, saying the presence of important investors in some companies allowed to advance the cause of blacks. The design of applied management cannot then ignore the organizational community and its relationship with the society in which it is immersed. Regarding the value of leadership, it must be stressed that an approach that conflicts with leaders of values is unsustainable. As they play a key role in achieving the goals, they would not have the energy to defend it. The example of a company that launched supercomputers, is this interesting title. The growth of this company needed to pay attention more and more to issues of markets and marketing. The founder could then identify with what had become of his company and preferred to move away, leaving room for a CEO whose sensitivity and values allowed to maintain a balance between the needs of researchers and those of the market. Leaders of values can act in every way.

They can be a considerable source of energy, including the implementation of optimization mechanisms. But they can also be the source of significant challenges and a denial of reality that can be damaging to the company. In all cases, the leaders of values cannot be neglected in assessing the choices the company makes or should make. It often mentions the importance of societal values of the leaders of beauty care companies. These values, including respect for and protection of the environment, are reflected in the nature of the ingredients used for the production, in the type of marketing in the recruitment and, of course, in the behavior of employees and franchisees with customers. It also mentions often the values of initiative and creativity of some leaders in the development of their businesses. But values do not always act so as visible or as spectacular. However, they always color the perceptions of leaders and change significantly their analyses, assessments, and worldview. They can act as blinders or, conversely, as warners. This is why the analysis leading to the optimal decision often has to take a break to allow managers to understand the beliefs and values that drive and compatibility between them and the established operational choices.

In this respect, the demographic characteristics of leaders are often in close interaction with their values. Thus, age, work experience, social origin, nature and duration of education, psychological characteristics, etc., affect very substantially on behavior. Some research has suggested that a leader who has lived changing experiences will tend not to undertake major changes to new or radical. These factors raise the question of the design optimization of business processes. When the elements of the analysis are available, their combination allows to design the operational objectives. The opportunities and threats, capabilities, values of leaders, and their concern for social contributions are the ingredients with which they must formulate a purpose powerful enough to serve as a guide to action for the members of the company and to optimize the competitive advantage.

The combination of ingredients is not a mechanical exercise: It is the heart of strategic thinking and can lead to a large number of choices. Each choice can thus be considered unique. Even when the business context is the same for all companies, the choices can vary depending on other factors considered. This also explains the fact that when the capacities are changed significantly, as at the time of a major acquisition or merger, or when leaders are replaced, the mechanism is very often re-evaluated. The combination that leads to optimization of business is an act of artistic nature. The ingredients needed to carry out the work that is available, but their use will produce a unique work, a bit like the realization of an artist's painting. We do not really know in advance if the table will be a masterpiece or a multicolored cloth.

In general, the decision-making approach must be compatible with each of the findings of the analysis, and particularly, with the contextual nature of the business, especially with the opportunities and threats it contains. The choice of a field of activity (key mechanism) should also be compatible with the company's capabilities. Finally, the choice must also be reconciled with the values of leadership and the desired type of social contribution. One would think that since the choices are unique, you cannot really learn from the experience of others. It's not quite the case. The companies' optimization of action shows us that those who do well do things the same way. That is where we turn now.

On rules, note that experience shows that some practices are the very essence of applied management. These practices may be set out in 4 rules:

- It must be different and unique. This means that in the choice of areas and objectives, it is important that the company defines a sufficiently distinctive way for its members, as its customers are able to recognize it. The difference, when perceived by the customer allows the organization to protect itself against the competition.
- To carry, use his strength (popularized as saying stick to the knitting). This seems obvious but the simple things are often taken

for granted and replaced with buildings which, although they are exciting for the players concerned, may expose the company to adversity instead of building on its most foundations solid.

- It should concentrate its resources in areas where we have an advantage over the competition. This applies especially when in many areas. The allocation of resources must avoid dispersion and strengthen competitive advantage. Thus, optimization case, the saying "do not put all your eggs in one basket" has validity only when resources are more important than what is needed to strengthen key areas, who are crucial to the long-term health of the company.

- We must choose the range of products as closely as possible, consistent with available resources and market requirements. This rule complements the previous one. We should be in more than one sector if one has surplus resources or operational considerations dictate. Thus, to use the example of oil, there was a time when to be in the refining, you had to be in oil production because it was the only way to ensure supplies to refineries.

We have noticed, design optimization measures can be dominated by the context of business by now. In this case, all the elements of the analysis are submitted to the deterministic nature of the analysis of the industrial economy. Resources are used to position itself in a world that is already fully established and which requires all choices. Only choices consistent with the context of business (especially economic) are considered. This approach tends to neglect the role of a leader and that of other corporate stakeholders. This perspective leads naturally to generic mechanisms, such as those that have been mentioned above and, in fact, deny, though cautiously, the possibility of truly unique solutions. The design of mechanisms can also be dominated by skills and company resources. This perspective is then more proactive since it considers such the context of the business as a corporate building.

This is a deliberately oriented perspective a future that must be imagined and created rather than suffered. Of course, the leaders of place here. A striking example of this approach is that of the company which is subjected to competition a barrage of innovations. For it, the economic climate does not really exist. It is always in the future and you have to invent constantly.

It should be noted that companies evolve and evolution seems to follow recognizable paths. Thus, stage 1, the company is simple, with a single product or a single product, no formalized and managed directly by the owner, who fulfills all the managerial functions, without a systematic approach to me, sour or control performance. In stage II, the company has grown enough to warrant greater specialization and the emergence of functions. Coordination is crucial and is assured by both formalization and a greater systematization and centralization of coordination tasks at the top. In particular, the assessment of the performance of officials is more formal and based on the achievement of operational targets agreed with the general direction. Planning is often the preferred management tool. Generally, the office of President becomes more important, which helps to ensure the necessary coordination, particularly by managing multiple systems in place.

If the company continues to grow, it undertakes new and diversi-fied activities that require a more decentralized organization based on product–market relationships rather than functional. This is stage III. The formalization is always important but on different grounds. Managers are evaluated on the profits they make in the areas that concern them, with a margin of maneuvers established in advance. Often each division operates as a business of stage II. In firms made in stage III, the trend is a decrease in the number of people working in the office of the president. The main tasks carried out there related to the financial management of all and the constant clarification of the rules of the game and the purpose of the busi-ness. The business life cycle is based on the management challenges facing the company as it grows and diversifies, five steps, and the transition from one to the other requires the resolution of a real crisis. Each of these crises can destroy the business.

After a so-called growth "by creativity," the company experienced its first "leadership" crisis. The second phase of growth, "by direction," leading to a "crisis of autonomy," when this crisis is resolved, growth continues with "delegation" and this leads to a crisis of "bureaucratiza-tion." The phase of "collaboration" may lead to a new crisis resulting from the multiplication of conflicts engendered by the democratic nature of this phase. This crisis should lead to a fifth phase.

This historical look shows us that the path in terms of optimizing management of this company regularities that manifest throughout his life. Understanding the dynamics underlying the evolution of the company allows to recognize problems that shift between the operational approaches and create the reality, and to better assess when these shifts become large enough to warrant a change in mechanisms and approach. The appreciation offsets bring us to the question of the evaluation of optimization mechanisms. This is particularly important when you want to enjoy the practices that the company has followed for several years or that of competitors. We propose some benchmarks to facilitate this assessment. So, assess the quality of the applied management must be a major concern of leaders, especially when the organization becomes complex and that they cannot directly participate in the analysis and reflection in all areas. We must, therefore, have criteria to tell if the direction is good or bad. The quality of business optimization is to harmonize the decisions taken in order to converge the efforts. This convergence idea is also a sense of consistency. One can even say that optimization is synonymous with consistency. The evaluation criteria used so extensively the idea of consistency. Four criteria can be used a priori three criteria can be used when the mechanism has been implemented (ex-post).

To make the evaluation a priori, it is to check if the operational approach as formulated is really based on the conclusion of the operational analysis. Hence the following questions: Is the chosen option or with the results consistent with the results of the analysis of the business context? As we have seen, the context of business generates opportunities and threats. Are they considered in the choices made? The mechanism he takes particular advantage of the opportunities available in context? Does it allow to deal with the most serious threats? We could also, for the assessment, to be more precise in defining the elements of the context of business with which you want to check the consistency.

So, to the question of whether the chosen optimization mechanisms are consistent (or consistent) with the results of the analysis of resources and internal capabilities, we must say that the analysis of internal capabilities clarifies what can be considered strengths or weaknesses in comparison with those of the competition. The tools normally have to be built on strengths. In some cases, it may be appropriate to work to reduce the weaknesses, especially when they may jeopardize the company, but most often the most informed choices include strengthening the forces and use them in the competitive struggle. Are the mechanisms built on the company's strengths? Do they take into account the weaknesses formulated? It could be even more specific, with an enhanced understanding of the value chain.

Thus, one might wonder if the solution allows using the resources that are available and which are not at work or if it strengthens the relationship between the value-creating activities.

When asked whether it is chosen option consistent with the desired social contribution, it will ask to what extent the company's social concerns are being taken into consideration by the applied management. The choice will impact what is valued by the company's members? Is it consistent with the leaders of values? It must be said that hindsight, we can assess the effects of management mechanisms to be short-term, such as profit for companies or long-term, as the clarity of the purpose for members and force the advantage competitive realized. Hence the questions: They confirm the company's short-term results of the validity of underlying mechanisms? Within these results, there is economic performance, but also social performance. Competitive advantage is also measured by indicators that are associated with the longer-term performance that include the cost advantage, product differentiation, brand differentiation and, in general, the establishment of barriers to entry higher. The competitive advantage can also be assessed by comparing the company with its competitors (quality, R&D costs, etc.). This calibration effort is called benchmarking. Finally, the mechanism there is an effective and flexible action guide for all staff and, in particular, for key leaders? The purpose of business is particularly important for the concentration of efforts. If it is too general, it is valid for all organizations and loses its grip on the members of the organization. If it is too precise, it does not leave them enough space for them to enrich.

CHAPTER 9

Operational Decision-Making

Many researches are deadlocked on the process of making business decisions and are interested only in diagnostic procedures and control. Admittedly, the neoclassical economic analysis offers solutions that perfect information. This shortcoming, involving the use of models (either very complicated or overly simplistic and therefore largely unused), was of little importance as long as strategic management was only relevant to very large companies (in which procedures played a decisive role). But the explosion of small business development has brought to the fore the role of the leader in operational decision-making. As a result, scientific research on entrepreneurship has been boosted.

Thus, decision theory has boomed and led to the concepts and classifications that are widely in use. In particular, decision theory has focused on the nature of the operational approach. But beyond the contribution of economists who seek to optimize, the major breakthrough came from sociologists, who are interested in decision-making in large organizations, particularly in big business. Nowadays, the role of psychologists, including specialist knowledge (knowledge engineers), appeared to be very important for understanding the mental processes of perception of problems, learning, and decision-making choice. After first showing interest in buyer behavior, they applied their models to the individual who makes an operational decision, namely, the owner-manager.

We specify, at first, the nature and type of decision, and especially the business decision. Second, we will consider its role in large organizations. Finally, we will show that, in small organizations, operational decision is for the leader himself. He returned to Herbert Simon for having distinguished three types of decisions we need to take. In the case of programed decisions, the problem to be solved is often well delineated and defined. It has all the information needed to reach a solution. For this, we use a rational model, logic, which gives the optimal solution, all things being equal. So we proceed in a sequence "BMI"—Intelligence of the problem, Modeling of the problem, and Optimal choice. This type of decision, saying programed, is reflected in the current operations of the company. These repetitive decisions, triggered by simple stimuli, require little complex information to achieve a precise choice. For such decisions, the computer can replace the operator in most cases (e.g., an instrumental banking operation). Their rationality is, however, more instrumental than logical to arrive at a precise choice.

Semiprogramed decisions, intermediate types are those that meet frequently with corporate executives. Suppose, for example, whether it is negotiating a purchase from a supplier: We must gather information on suppliers, prices, quality, deadlines, and so on. This information is more or less reliable, and more or less easy to obtain. It requires to specify what exactly is sought. We need to structure this information, which means that we have procedures or even analysis grids more or less precise, logical. We must choose between alternatives, based on sufficiently relevant decision criteria, but be assured that we took the "best" decision. Information systems specialists strive to develop expert systems (e.g., for medical diagnosis, for financial analysis), or support systems to the decision, which aims to provide sorts algorithms to implement the decision-making process.

Many tools and management techniques in the various functions of the company are in fact of decision support tools, rather than, as too many students believe, tools that directly give the decision. In particular, in the semiprogramed decisions, poorly structured, the role of judgment, often leaning on past experience, is very important. Some of these decisions are repetitive enough, sufficiently precise about the nature of choice, and properly supplied with relevant information to move toward programing. However, some remain too uncertain and too complex, and are close to nonprogramable decisions stronghold of operational action.

Nonprogramable decisions have the following characteristics: They have a large degree of uncertainty—what information would be required to make a "logical" decision "rational," "optimal," are either too few or too many, be biased or simply unobtainable because they affect the future and take account of others. They have most often a high degree of complexity, since many variables come into play, so it is not possible to rely on a simple model, linear, deterministic (type "A is the cause of B"). They exhibit a high degree of indecision in the nature of the problem.

Often the question is to find what the main problem is before questioning the choice itself (the Anglo-Saxons talk about the search process).

As a result, this type of decision is based on the thought process of the decision maker; he is the one who will choose the relevant information, decide the situation, identify problems, and feel the choices that seem appropriate, with its own mental patterns. It involves a major role of intuition; it is a mental attitude, which makes it "feel" that such a decision, solution, and so on is "good," "right," "satisfactory," and so on. Intuition is based on own decision maker characteristics, some saying they are innate (the "flair" to own some makers), others acquired (experienced manager). Mintzberg took over the distinction—moreover, scientifically controversial between the right brain (part of the sensibility and intuition) and the left (part of rationality and logic). Of course, most of our decisions are "bounded rationality" and are the result of a mix of logic and intuition.

But above all, Mintzberg, studying the decisions taken daily by company executives or any organization, showed that the overwhelming majority of them were unstructured, largely based on the "intuition" of the decision maker, that is to say, not justified by the use of a model or logical-mathematical demonstration. In many cases, the "models," "standards," and "technical" management are used to justify the decision intuitively. In desperation, the decision maker can be justified in a model "irrational," that is to say, not scientifically proven. The cartoon is the use of graphology, numerology, and other "para sciences" for recruitment, or the use of astrologers from key decision makers. But one wonders if the number of "models" often sold dearly, including small businesses are not mere means to justify "scientifically" a strategic decision that it is impossible to "prove" the "truth," and let alone optimality (discontinuation of a business on behalf of a strategic matrix, for example). Finally, it involves a complex process, due to trial and error, progressive centering on the problem, flashbacks (to get information or restate the problem). This process should largely to learning: the decision itself, then the sequence of decisions made by the leader, who forges his own mental processes, and even its own clues and its own grids, even informally.

In total, the soft decisions are based on substantive rationality: They are based on linear relationships of cause and effect; they lead to an optimal solution, logically demonstrated. Nonprogramable decisions are based on limited rationality in an individual, in a procedural organization. The solution which leads the result of "deliberation" (negotiation in an organization, evaluation of the "for" and "against" in an individual); there is no "proof" of the validity of the chosen solution, only a "justification" after a heuristic approach, turned as much on the research of the problem as its solution, which is merely "satisfactory," and not "performed." So, we developed new logical-mathematical ways of thinking to try to decipher situations largely marred by uncertainty, faced with ignorance of the behavior of the enemy. The "game theory" was then applied to the economic analysis of the operational approach of shareholders in competition. Faced with a situation of interdependence (your score is determined by the choice of the opponent, or a "state of nature," the fashion next season, for example), game theory helps locate issues of a strategic choice by the use of decision criteria that reflect the attitude to risk.

But they do not give the single optimal solution. Game theory allows us to situate the issues of a strategic decision by the use of decision criteria that reflect the attitude to risk.

For example, suppose a manufacturer of women's clothing industry articles wonders what collection it must start knowing that its success is linked to the mode that prevails at launch. The manufacturer can adopt safe behavior: He will choose the collection, which, regardless of the mode, give him the minimum maximum profit. The risk of loss is reduced, corresponding to cautious behavior. This criterion is called optimal. But the industry can reason in a more "rational" and strive to choose the solution that had he been aware of fashion, give him the minimum of "regret." For this, it calculates deviations from the best solution (the most profitable collection) for each mode, which gives the matrix regrets. We then look at the option that gives the lowest maximum deviation from the "best" solution (economists say the maximum opportunity cost lowest).

This solution is called "minimax regret." In fact, there are many other criteria that may be used from this matrix. The important thing is not in the crude solution, but in the thinking imposed by mounting the matrix on the choice of operational variables and assumptions about the states of nature. It would add the actions and possible reactions of competitors, the probability of occurrence of each of the modes. In short, we are limited rationality, and this type of matrix can only be a tool for reflection in a heuristic approach. It must be understood that it is the same for models and analytical frameworks that were presented: They cannot under any circumstances replace the strategist own reflection; rather, they should lead him to ask issues (or even question) by helping to raise issues. Say they give "the" solution would report quackery!

All current operational analysis formalized, focusing on game theory, was multiplying situations and decision criteria. The concrete contribution, however, remains very disappointing. The authors who worked on decision-making, including operational decisions, were taken to distinguish this problem depending on the size of the organization. The most developed analyses concern big business. But there is a growing interest in the study of decision-making in small organizations. This is two orders of different strong concerns. Indeed, in large organizations, one can say that the decision is largely based on procedures and interpersonal and collective relations. The rationality of the decision is said, according to Simon, "procedural." On the other hand, in small organizations, decision-making is the act of an individual, even if he is surrounded by advice. It is therefore more a mental process, and we must speak of rationality "limited," according to Simon.

In large organizations, it was seen that the trend was the differentiation of tasks and functions; vertical and horizontal integration is achieved through techniques and procedures to increase motivation and raise the morale of "organization. The decentralization of decisions, including decisions about the "business optimization" at the level of product-market divisions participated in this double movement of differentiation and integration. In big business, we will prioritize decisions according to two major characteristics.

The complexity of the problem raised. More key variables (the elements of the problem) are easy to identify and quantify, the more they are connected linearly them over the problem appears simple to make. In contrast, the more the variables are difficult to spot, are qualitative in nature, and are interactive, the more the formulation can be described as complex. Advances in management techniques tended to division problems, so as to simplify their wording to make decisions "operational."

The degree of certainty and uncertainty in the nature of the choices to be made, the type of decision. Some decisions relate to a specific choice (decide to do or not to do, decide when, how much, etc.). Other decisions are based on inaccurate choice, broader, more nuanced, and may even involve asking first what the question to resolve is.

It then leads to several types of decisions within the structure. There are the problems mentioned, dedicated to operational planning. But analysis of the decision process is complicated in decentralized companies, since a part of the strategic subsystem (complex decisions with choices specify) is "down" in the hierarchy of the organization. This results in important consequences.

To successfully master the operation of the organization, it must privilege the "simple" and "precise." For this, establish rules as simple as possible, which will be formalized in terms of specific procedures. For example, any investment project will be adopted only if the rate of return, calculated according to predetermined rules, is greater than a figure floor. Similarly, the development of activity will be reduced to a target goal of market share or turnover. The more complex problems, such as "quality" or "moral" will be simplified with the help of indicators.

This approach introduces a procedural rationality to large hierarchical organizations. Individuals and subgroups (divisions, departments, etc.) will endeavor to influence the determination of rules to benefit (called "internal operational approach"). A major concern will be to provide rules which, while seeking the goals of the organization (efficiency), through a better operation, better use of resources (efficiency), do not cause any dissatisfaction and major internal conflicts (effectiveness).

In total, this is to work toward a satisfactory solution, that is to say, which maintains the stability of the organization, while ensuring its sustainability or competitiveness of capitalist enterprise:

In reality, the procedures for decision-making are not as simple and unique as in the operational planning manuals. The researchers showed that for operational decisions, the solution gradually emerged is "modeled" according to the multiple influences that were exerted on decision makers, with possible back and forth, or trial and error. Large public projects (Big Library, the Channel Tunnel, etc.) illustrate this approach of "modeling." It is very common in very large organizations (large industrial groups) subject to multiple pressures. Some authors even speak of "garbage model," saying that strategic decision reflects numerous influences, not just the economic rationality of optimal allocation of resources and maximum profitability. We will talk about broader rationality, to express the idea that the logical choice is diversified, in an "irrationality" apparent.

The structure of large productive organizations reflects a desire to master this complexity and prioritize the types of decisions. Most often, they take the form of group. The group will be defined as an integrated set of companies.

The integration is primarily financial. The "head" of the group consists of one (or) holding company that owns a portfolio of investments in the capital of companies in the group. The first set of entries corresponding to the core of the group comprises subsidiaries and majority control companies—typically more than 66% or employee controlled by the set of successive participations. The second group includes companies financially integrated, but without majority control (often as a result of acquisitions and acquisitions). Finally, the third set includes joint ventures, and sharing is done between several groups, after alliance operations.

The integration is then industry: There is one area that consists of almost integrated companies, that is to say very dependent on one or more companies belonging to the group. The management and operational decisions of these firms are closely controlled by the group.

The financial and industrial dimensions can be combined by the interpenetration of capital, industrial, and financial groups which were then lead to hypergroups, real nebulae companies, and subsidiaries.

Within these groups are three hierarchical levels, through the information and decision system: finalization, entertainment-control, and operationalization. The essential problem of the strategic information system is to circulate relevant information up and down (considering the cost information) to quickly take the appropriate decisions, particularly to change the operational approach with maximum speed and minimum cost at large. This imperative is even stronger than the context of case is turbulent. The larger the organization is small, the more we find the following features.

The decision is largely attributable to the entrepreneur. Although he is surrounded by advice (family, accountants, bankers, etc.), he alone is responsible for decision-making and execution.

The company is highly dependent, in the broad sense of its environment. It will often be more difficult to have a completely independent approach. The entrepreneur will pay attention to messages from their environment, which will trigger operational responses.

The structure is not formalized, little hierarchy. Information and control systems are closely linked to the personality of the leader.

The three levels mentioned previously are completely intertwined: An operational decision may have operational consequences (the choice of new material can result in a change of suppliers, customers, etc.), but is not necessarily perceived by the officer in "immediate."

The decision process is limited information; we are Simonian worlds of bounded rationality. The goal is not to make the "best" decision but to identify the key issues, collect a "reasonable" solution "satisfactory" to justify (usually from its own mental models), and then implement it. This process of research, much of the problem as its solution is heuristic: the process is gradual, tentative, made of trial and error, learning-based decision makers and his cognitive ability (ability to analyze much to synthesize, logically deduce that induce intuitively).

Two major categories of processes it is customary to distinguish:

The reactive and proactive processes. The reactive process results in a response to a stimulus (something new in the environment). The proactive process is to create this development, especially through innovation, through more aggressive than defensive, and so on.

- Emerging and deliberate process. The result of a deliberate process plan, intention, of a clearly stated vision for a certain duration. The emerging process ("incremental") is the result of a gradual adaptation to changing conditions or stimuli (the change may have been triggered by the company itself).

We cannot say a priori what the best process—even if for a long time management literature favored the proactive and deliberate process, that is to say planned. In fact, many factors are involved:

- The nature of the decision: Buying an expensive machine will be a rather deliberate proactive type adopting a substantial spontaneous order reactive type emerges.
- The nature of the type of activity and the environment: The environment is more turbulent and complex; the contractor will adopt an emerging reactive attitude.
- The nature of the entrepreneur and his aspirations: The entrepreneur seeking growth will not have the same attitude as the one who seeks the sustainability and survival of his business.
- The nature of the type of organization: Structures (mechanistic) will be better suited to proactive decisions adhocracy emerging reactive decisions.

In total, it is important, finally, to note that small businesses are increasingly conditioned in their operational decisions by their inclusion within a network of companies and institutions. Finally, the small business can grow by structuring in the form of group. This will include a company managing the investments (the leader of the family shareholders) in various companies in the same industry or in different sectors or corresponding to business functions (purchasing, production, and marketing design). In contrast to hypergroups, we speak of hypogroups.

CHAPTER 10

Applied Operational Analysis

It is appropriate here to ask how to drive an applied operational analysis. A number of observations are in order. First, we must distinguish between the operational diagnoses of the recommended operational decisions. Analysis of a case of optimization does not tend to search for "the" solution, but should focus on the detection of problems and the development of their interaction within the operating system. We must not forget that, in concrete situations, it rarely leads to a single decision, precise, deliberate, and definitive. The importance often lies in the awareness of the issues from the executive. Indeed, solving a problem often depends on the setting relations several operational variables, and this interaction is likely to lead to other problems; solving a problem is a process that takes time, which is widely emerging, with trial and error. This implies monitoring over time, support, and likely will lead to question some options.

In this perspective, it seems naive enough to think that the solution must flow logically from the diagnosis. As we have mentioned many times, the instrumental rationality of attitude stems from what applied operational analysis was first applied to large diversified companies, usually in the consumer goods sector unmarked, in a climate of regular global demand growth, on stable markets, with a renewal of controlled products. This approach reveals weaknesses when the company is vulnerable to a complex and turbulent environment becomes, or its small size. The growing reluctance to solutions "logical" also comes from feeling increasingly asserted that the company's competitiveness stems.

Second, keep in mind what we said in the introduction. The difficulty of the applied operational analysis is the fact that it requires a mastery of both concepts, which, moreover, are not always clear, and management practices. Let us recall the skepticism displayed by some vis-à-vis authors of the operational approach of education, and especially the case method, for students who have not yet known the decision-making in a complex organization. However, it may be observed that by careful use of cases from reality, the student can thus bridge the gap between the concepts, tools, analytical frameworks, and highlighted operational problems of business. It may thus raise the complexity of the process, aware of the interaction of phenomena. It is, however, dangerous to let him believe that the grids and other tools provide "the" solution, as would imply a clumsy operation of certain models: SWOT (Strengths, Weaknesses, Opportunities, Threats), BCG (Boston Consulting Group), value chain, and so on.

We then see what is the role of the tools they used to decipher a complex situation to reposition the issues and to frame solutions. These are, according to the heuristic method of aid instruments in the decision. Often, they allow to highlight the gaps, particularly in terms of information: The student is often surprised to find very little quantitative information, and if there are (balance sheets, financial accounts) hastens to analyze, to the detriment of a more comprehensive synthesis. It is worth remembering that, in reality, the decision maker has only a very partial information, especially quality ("good", "bad", etc.), subjective (perceptual), and relative ("better" or "worse"). Furthermore, the encrypted information is retrospective, or static or instantaneous. One of the pitfalls commonly encountered by beginners is to stick to the problems within the company, without looking at internal strengths (which makes it competitive) and the changing environment of business (which poses the problem of positioning), content, often a critical organizational diagnosis, which is not the subject of applied operational analysis.

It is clear that the combination of reflection, supported the handling of concepts and tools, and action, resulting from a decision lucid awareness, realistic, pragmatic problems facing the company, requires learning analysis applied. Gradually, as we carry out this type of approach, Simon shows that the mental patterns we forge, problem-solving methods (much like the chess player) and, previously, detect problems. Therefore, the operational approach consulting tends to forge its own analysis grids. It can be difficult to teach them to others, who probably will not have the same mental patterns. But the more you work in the field of "nonprogramable," the "poorly structured" should be more methodical, in the sense that it is necessary to be aware of the approach that we adopt: Again, the chess player is a good example. It is clear that good mastery of concepts and tools taught in applied management will be valuable, knowing he should use them with caution.

The applied operational analysis must adapt to the circumstances, including contingencies: size, organizational structure, relationship with the environment, industry, and so on. Obviously, the most important contingent variable is the size of the organization. Clearly, the applied operational analysis arises in the same terms in M form organizations, as in small businesses. In large firms, one can easily distinguish the operational approach of general political activity. In small companies, we highlighted the strong interaction of issues and levels. That is why we must speak of a specific, widely emphasized by the authors of the operational approach in small to mid-size enterprises (SMEs).

We should also mention the difference of problems according to business segments: Moreover, organizational adaptation boards are often specialized, because you have to know the activities and markets. Here too, we should speak of a specificity of the problems and methods of analysis (for example in agriculture or in services), which have been neglected by the most common theories (still very much attached to the mass goods of the second industrial generation). The method we propose will result from the approach that applied operational analysis shall:

- beyond logic, a highly instrumental rationality, method-based procedures, culminating in a heuristic decision, based on limited rationality, process-oriented, and organizational dynamics;
- spend an analytical and linear method, culminating in a holistic approach, integrative, systemic, which takes into account the interaction problems, promotes the return on the approach or past results or trial and error, involving lead on solutions that will have meaning only if they are accepted, integrated and implemented by the decision maker (which is far from certain).

The approach is based on four pillars: purpose, organization, environment, activity. It distinguishes between the corporate level and the business level. Let us repeat that the smaller the company, the more these levels merge. The corporate level is based on the major relationships between the pillars: vision, legitimacy, culture, which broadly express the issues raised by the values, aspirations, of the leaders and owners, of the Company and of the members of the organization. The last relationship evokes the couple image (the company as seen by the business context) and identity (the organization as it sees itself) leading to appropriate communication approaches.The level "business decision" is based primarily on the couple trade-mission which causes problems of identification of the relationship between competitive positioning and competitive advantage. The goal-related activities are reflected in planning, whether explicit, procedural, or not explicit, procedural (vision).

Finally, analysis of activities poses increasingly the border problems of the organization, in terms of external or internal transactions. Note that the arrows are in both directions: We are dealing with a system, complex, open to its environment, finalized, and should be regulated. The essential problem is to arrive to ensure a dynamic coherence to this system; the one exceeds the identity of the sum of its parts. However, each of these is likely to change continuously, abruptly, or continuously. This implies a constant operational monitoring, even though in many cases the changes are imperceptible, emerging, until they produce ruptures, "catastrophes" in the system. During the analysis, and on the proposals, it is important to reflect this dynamic interdependence. At that time, one quickly becomes aware of the complexity, the difficulty in predicting the consequences: We talk instead of "practical solutions" or "emerging."

There are several types of cases. The most common type is to present the problems of a large company, often focused on "business decision," with, supporting, much information about the evolution of the sector, the figures available on business, markets, and competition. Note that when the company is well known, the analysis may be biased by knowledge of the choices the company has actually adopted. For example, if the company very interesting to analyze the diversification mechanisms, is biased by knowledge of the products of the firm, even meteorology (no snow) or course (unexpected) of the dollar in recent years. Therefore, it may be more interesting to take a company of any size in less publicized areas. The advantage is when one does not know the operational approach effectively implemented and what happened to it.

But, again, the use of a case involving a small business has the advantage to reveal the interdependence of all variables and all levels of analysis. Therefore, it is desirable to provide cases involving the small and medium enterprise, allowing to practice the applied operational analysis. Moreover, these cases provide only limited information, which brings us closer to real situations and avoids the pitfalls of management diagnosis.

Mr. and Mrs. CQCI are farmers. They rear livestock for milk production, which was their main resource. They then announced the limitation of production of milk, and milk quotas per farm are set. The couple is being brutally forced to find another source of income. The problem is particularly serious that they are not original farmers, but, on the other hand, they feel freer to try a new experience, and even to stop operations and leave as employees. Like many farmers, they will grope. They begin to question (chicken? rabbit? mink? duck?). And they end up opting for breeding ducks. They inform about the conditions of breeding, "read books," and so on, and, after many experiments, begin to make Foie-Gras. They decide to specialize in liver upscale fat based on a clientele of top restaurants in their area, who wish to find nearby producers could make their Foie-Gras corresponding exactly to what they need for their kitchen. But the Foie-Gras produced in the region is of excellent quality. Their problem was then to develop highly personalized relationships, including in the product, and then develop a communication showing that it was possible to manufacture a high-range Foie-Gras and, accordingly, a regional product, tourism, in a region rather marked by industrial decline, and wishing itself change its image. They experience first successful: Turnover has doubled, but relies on the half duck. It took two recruiting employees to care for the livestock.

You learn quickly in business optimization mechanism that when all goes well, the problems are not far away. Indeed, demand (too) responded perfectly, and this micro-enterprise is in crisis growth. Leaders do not have the money, so they want to ensure the future of their two children on the farm. Moreover, the risk of loss of quality is also important, as restaurateurs still very demanding (the product is not standardized and must constantly monitor its quality), especially during peak periods (holidays). It follows that certain sales cannot be concluded.

Moreover, the company filed trademark; it begins to produce and distribute under its brand, besides fatty liver, other specialties from duck, but it could diversify into other gastronomic products. An opportunity presented itself: A Foie-Gras producer fee from another region is interested in their production and offered them to participate in the expansion, acquiring their skills. But leaders are wondering how to keep their identity, be sure about the quality, not to lose its independence, while ensuring the expansion. This small case has several interests. First, it shows that in any organization, however small it may be, whatever the sector and type of activity, may arise optimization of business problems. Then we see that in its apparent simplicity hides complex actually strong problems. First is identifying the intention of the owner-managers: We feel any contradiction, since they seem to pursue both growth but also sustainability (secure the future of their children) and independence; they seem now too committed to exit this activity duck.

But, as so often, they will no doubt lead to prioritize their aspirations. Second, note that their legitimacy is strong: regional high-quality product for a high-end customer in a region that seeks to change his image and wants to develop green tourism. Perhaps they could use this to get regional support (if they are already). In third place, it does not seem to be a conflict between the spouses. It would be interesting to know the intentions of the children, and especially how they will establish any relations with another producer. Indeed, we need to know what the nature of the legal relationship: How will operate the know-how? How will the quality controlled, which we have seen that it was not standard? Who will establish business relationships with restaurateurs, with private clients? Who will own the label, the brand? Which we have seen that it was not standard?

The other option would be that of diversification, ongoing, to other high-end products, benefiting from the notoriety. But the leaders they have the know-how? Will he not have to, again, subcontract, with the same control problems? Finally, eventually, does he need not increase capital, with new shareholders? In this case, the owners actually experienced some problems due to uncontrolled growth and some difficulties in outsourcing. But the most important lesson of this little illustrative example, the need for consistency between all operational variables.

The grid that is proposed, the result of work conducted within ExpertActions ExiGlobal Group (group of experts and consultants in operational approach). The implementation of an operational action plan includes two stages: The first stage consists of preparing the plan, and the second step consists of monitoring the action plan. The first step is very complex. First, we must distinguish, as the initiative comes from the entrepreneur who encounters an operational problem, or consultant, which is a lack of operational approach taken, clear or even an absence of coherence. It should also be distinguished according to whether the approach is reactive, linked to a problem or an opportunity, or proactive: deliberate approach of the entrepreneur who wants to flatten its problems, or planning major operational decisions (transfer, acquisition, transmission, etc.).

Must also be distinguished according to whether the initiative comes from the entrepreneur who encounters an operational problem, or consultant, who notes the absence of operational approach taken, clear, or even a lack of consistency.

(1) In any case, proceed to a presentation of the "operating system" of the company, with the help of the analytical framework that we presented above. This step involves the participation of the entrepreneur, who must not only find information but also explicit representations (e.g., how it perceives its environment, its position, its distinctive advantages, etc.). Generally, this essential phase is long, because the decision maker realizes that rationality is limited and must strive to clarify its representations. The back and forth should allow a gradual appropriation of the approach. This appropriation results in the presentation of his vision of the operating system for the coming years, based on what needs to change (goals, organization, activities, and business context). This step is crucial, because it implies that the decision maker is aware of critical business issues for his business, and he is able to lead to major business lines. This can cause a real challenge to its choice and its logic of action, so that the consultant should adopt a participatory attitude of support, without imposing anything, and avoiding value judgments.

(2) We can then draw the first draft of an operational action plan, which should at this point be evaluated on two levels:

First level: What relevance? One must wonder if the options are mutually consistent (big investments, but not debt), if feasible (adoption of sophisticated technology), if they solve real business problems (too narrow market, for example), if they do not cause conflicts (creation of new functions), if they are not too risky (highly innovative product), and, last but not least, if they are realistic (especially terms of expected outcomes, or calendar to keep). This evaluation can be internal (with the help of the decision maker) or external (with the help of external experts and possibly neutral). It is important to know the reactions of the decision maker (family, employees, accountant, banker, etc.) if it is an SME.

If relevance does not appear sufficient, we must then "looping back" and question some operational options.

Second level: What performance? It should measure the cost of the proposed operational decisions and outcomes. The cost should be understood in a broad sense, since it must include all the problems caused by the change (abandonment of goods and resources, depreciation capabilities or qualifications, resistors, etc.). We must also take into account all expenses incurred (a new machine will involve training expenses, for example): The intangible expenditures prompted by a hardware investment can be substantial ... and are often funded with reluctance by bankers, who cannot make guarantees!

If the cost–benefit ratio proved unfavorable, we should challenge the action plan. In total, after successive iterations, it must result in a set of proposals, which lead to the second stage. The implementation of the operational action plan. The action plan includes, concretely, a number of operational decisions:

(1) They must be programed in time. It should adopt a logical sequence: For example, if you decide to develop new technologies, it will provide one of the workers training plans for them.

Similarly, if you decide to export, it will provide a training program for foreign languages (and probably more), and so on. Often this sequence into action is underestimated in the plans (e.g., the electronic booking process for some companies was implemented without sufficient training of employees), or with insufficient margins of freedom, making delays and catastrophic incidents. It should also include specific external intervention, coming to support certain actions (e.g., an export decision implies the use of guardian and expert networks).

(2) Once this sequence of operations has been programed, it is appropriate to monitor. To do this, the decision maker must have a board walk. It must first enable it to monitor the monitoring of the implementation within the organization, to identify delays, and to justify the cause.

But it is also to adapt to changing circumstances and often unpredictable. The decision maker must have warning indicators and perform a function vis-à-vis the environment before. The control function should be possible continuity over time: opposes that change is often gradual, incremental, or received late, on the occasion of incidents (defection of a major customer, for example). The role of the adviser or consultant then is to ensure both tracking and alert, with appropriate scoreboard. When disruptions occur, they can simply call into question the timeline involving a partial revision of the action plan. However, if they are more radical,

Thus, we see, finally, how the diagnosis phase is inseparable from the subsequent phase of implementation. This observation leads us to relativize the value of the conventional method of cases, too focused on the initial phase. The ideal is to follow the case up to the implementation of several years!

CHAPTER 11

Optimizing Operational Choices

Optimizing a company is in a sense to manage the change. All strategic management literature offer procedures and processes that allow the strategy to be adapted to maintain or improve the consistency between the organization and its environment. In most cases, this adaptation is done by gradual changes to the existing policy, changes that can be scheduled by management or that emerge from the action on the field. Similarly, the structure and systems are constantly being developed to strengthen the competitive advantage and effectiveness of the strategy.

Thus, an appropriate strategic management adjusts the organization continuously to prevent it from living crises, and thus avoid having to make a major change. In this case, change management comes down, essentially, to manage the company every day. But it happens that an organization is obliged, for a variety of reasons (breaking in context of business, internal crisis, prolonged inertia, etc.) to transform radically. In these rare but critical situations, the ability to manage a radical change becomes crucial. This is particularly important when the organization is complex, not least because of its size, the diversity of its activities and its geographic dispersion. Indeed, in a simple and small organization, the challenge is to define the new direction. The installation is relatively easy to control. However, in the case of a complex organization, beyond the difficulty of defining the new approach must bring all members of the organization to achieve in a context where the ambiguity and diversity of views and interests make it difficult to achieve a consensus. In these circumstances, it is often easier to destroy the existing organization than to build a new performing organization.

It is difficult to define a priori what a radical change is and determine when a strategic change can be considered radical. In fact, small changes can have significant consequences, and therefore precipitate a rupture. For example, a simple move can cause a cultural revolution that nobody had foreseen. Furthermore, a great transformation, announced with fanfare, may not produce the desired effects. Thus, despite the many attempts to reorganize GM, the automaker appears to be finding it difficult to change its ways. Also, what is a break for some may be only a change irrelevant to others? This is often a matter of interpretation, but that is of importance for the management of change. Despite these difficulties, the literature provides some useful guidelines to define the radical change. Starting from the idea that an organization is a configuration, that is, a coherent integration strategy, structure and culture, we can consider a radical change as a configuration change.

A radical change, we can also call processing, implies a new strategy (in terms of business model, positioning, development, etc.), which requires a new structure and a new culture to form a configuration consistent. This shift is based on a new framework, often developed by a new leader. Thus, the radical change would be cropping (Reger et al., 1994), while the continuous change would be a change within the existing framework. If the basic assumptions (core beliefs and values) are challenged, we could even talk about recreation. From this perspective, organizational changes would be made long periods of convergence around a relatively powerful configuration,

A fascinating example is the hospital system transformation project, just like what is done elsewhere. Traditionally, hospitals were characterized by vocational guidance-oriented medical specialties, and this model has always had great legitimacy. However, in recent years, it has undertaken a major project to transform hospitals to develop customer orientation, characterized among others by groups according to the needs of patients. Such a transformation requires a significant change in perspective for specialists who have always dominated in this environment. It goes without saying that the realization of this new strategy–cultivation–structure configuration is a long-term project.

From the most frequently reported in the literature cases, it is possible to establish a typology of the different forms of organizational transformation: the creation, revitalization, reorientation, and recovery. Recreation, or the change in the worldview of the organization, is the most profound transformation since beliefs about what the company and its purpose must be fundamentally changed. This change in perspective that redefines the organization to its environment relationship is inevitably accompanied by changes in values and practices, the nature of the field of activity, and structural arrangements. The rebuild is generally a proactive transformation that is to say when it anticipates crisis.

Indeed, the prospect of change associated with a recreation is generally perceived as brutal, because it is experienced by members of the organization as the destruction of what they know and cherish. It generally follows the arrival of a new management team, carrying a new vision. Changing the way of seeing the world often requires dramatic action to signal that it actually intends to carry it to term change. In particular, managers are often replaced abruptly, old traditions are disappearing, new symbols appear, new behaviors are valued and highlighted, etc.

This type of change is perhaps the most difficult, especially if the members of the organization are not convinced that an early crisis is inevitable. It requires extremely persuasive leadership, as people need to make an act of faith in a context where there is a lot of ambiguity. Beliefs are very difficult to change, because their replacement creates a lot of insecurity. Unlearning, which is inevitable, has important implications. Suddenly, members of the organization see more of the skills that made them proud become obsolete. Finally, the translation of the new world into action requires a very important learning, especially since usually everything is changed both beliefs and values, fields of activity, and operating rules.

The transformation of Hydro-Q is an excellent example of recreation. After a long period of glory when the corporation was recognized as important, thanks to its prowess in the development of facilities such as James Bay, its purpose has been questioned as economically (its debt and operating costs increasing significantly) as well as social (indigenous people, among others, criticizing his arrogance). Hydro-Q has been transformed: a dam builder oriented engineering; it has become a seller of electricity customer oriented. This shift was not made without difficulty, and it continued over a long period. This kind of change is so profound that few organizations dare undertake if they measured all the consequences and all the difficulties. It really is an entire revolution, and few companies are trying it.

Revitalization, or changing the organization of practices, involves something of a challenge to itself, rather than its worldview. This is a change of perspective about the potential and the expectations we have vis-à-vis the organization. Ultimately, in addition to the change in values, it is a change of business scope and structural arrangements. Revitalization is usually a proactive transformation. The performance of the organization is not catastrophic; the time allotted to perform better is enough. As the change does not affect, at least initially, the purpose of the organization and its relation to the environment, it is shallower than recreation. This is, however, a change in practices affecting the whole of the organization and, therefore it introduces disturbances in which digestion can take several years. It is generally perceived as less brutal at first, because it creates less insecurity and can even be above-mentioned in some of the enthusiasm for new challenges.

Revitalization does not always mean a change in leadership. In addition, leaders who initiate change, even if they are new, often come from within the company. This change, even if it is started at full speed to reduce the resistance, takes a long time for the upcoming reality because the required training can be considerable. Moreover, since it requires significant effort, and even more if we want to continue to improve results, the cruising speed is difficult to maintain. Jack Welch at GE is a good example of a change that has given a new life to an already successful company. By setting very high performance objectives ("to be number one or number two") and by setting up an organizational development program requirement, Jack Welch transformed the organization from within. This process has had a significant impact on the culture, structure and, by domino effect on the strategic positioning of the company.

This type of change is usually undertaken in established organizations. In general, increased competition accompanies or promotes this type of transformation. However, the competitive situation also reveals a significant potential that the company can, if it changes, leverage to differentiate them by exploiting new competitive advantages. When a company is considering a shift or a change in business area, the current activities do not seem to meet expectations for the development of the company. The field of organizational activity can then be extended or activities gradually replaced by new. In this transformation, it is primarily the relationship to the environment that needs to be rethought. Changing beliefs and structure is related to the evolution of the field of activity. Although the organization is not in crisis, the shift is generally perceived as legitimate by members, and is even often considered welcome.

While in some cases, transformation is related to the decline of the current sector, in others it rather belongs to an organization's growth logic that exceeds the capabilities of existing activities. In both situations, organizational changes can be very important to adjust the structure and culture of the new strategic positioning. However, as in the situation of decay, the magnitude of change is anticipated in the situation of diversification, it is often underestimated, as demonstrated by the case of DuPont. Although the change is extensive, there is opportunity to lead it gradually and then it appears less severe than other types of transformation that we have described above. However, learning new activities is usually more important than what had been expected, which may create a shock. Although resistors are less great to leaving, problems may arise along the way when the magnitude of the necessary change becomes more evident. This type of transformation can be done without changing the management team, especially when the strategic reorientation is designed as a natural evolution of the company.

The transformation of DuPont is an example of a painful reorientation, but successful. DuPont's diversification into new areas has led the company to create a new organizational configuration, multi-divisional form, which is now widespread. Closer to home, the company Gildan, who managed in a short time to become the North American leader in the manufacture and wholesale of T-shirts, has recently diversified. The company decided to expand its range of products (socks, sportswear, and underwear) to create its own brand and sell directly to retailers. It recently announced a first down from the anticipated results, among others, because the range of products available at retail was not adequate. Is this the beginning of the hard learning of a new craft that could possibly lead to a change of culture and structure? Finally, here too competition stimulates change. It is generally strong, involves the performance of the company in the chosen field, and forces the reconsideration of the choices that were made earlier.

The reorganization, or restructuring in the short-term survival, is an operation to rationalize the activities of an organization that is in a desperate situation. It is necessary to reduce costs substantially and to restore order in activities. Although possibly fault reviewed the strategy and culture, the focus initially is to make significant cuts to stop bleeding. Recovery is the form of transformation that usually comes to mind when talking about radical change is the response to a crisis. In this situation, the resources are not sufficient to ensure the normal operation. It is necessary to carry out emergency surgery to save the organization. The change is sudden and very painful for members of the organization, who live a situation of failure. However, evidence of the disaster shows that nobody needs to be convinced of the need for change.

The change must be done urgently. There is not much time for discussion and reflection. As mentioned, the structural arrangements are first changed dramatically and discontinuously. It was only after having ensured the survival of the organization that the strategic repositioning, the values and their beliefs become a concern for the leaders. Change, even if the dramatic, creates less resistance than recreation. Learning is often important, but perceived as less brutal, though painful, because it is, in many cases, desired by many.

Generally, leadership that leads to recovery is new, because the leaders in place do not have the legitimacy to achieve change, and are generally considered responsible for failure leading to the restatement. The rescue of Chrysler under Iacocca is the most celebrated probably adjustments. Indeed, the new president saved the company from certain death, especially by a cabal with the US Congress to ensure they have the necessary funds for the revival of the auto giant. Iacocca who knew the industry for having worked his entire career in a competitor has installed a new management team and implemented a plan that required serious cuts. But once the defeated structure and reduced costs, he managed to give a new positioning and a new culture at Chrysler, with innovative products that have enjoyed phenomenal success. It is the creation of this new configuration that has enabled the company to regain hope for the future.

In contrast, the case of Nortel Networks is testament to the fact that the restatement is not a matter of rationalization. The company, which continues to reduce its staff and activities, is unable to reposition and continues to experience significant performance problems. As we have seen, the radical change can take many forms. Depending on the context in which the business is located, there seems to be an appropriate type of transformation. Now that we know the main forms of radical change, what can we say about how to manage?

The literature on organizational change management is very abundant, but it is much less on the management of radical change. The few authors who discuss generally propose a model in a few steps that describe the challenges at each stage and how to cope. Thus, Allaire and Firsirotu (1985) emphasize the Cultural Revolution associated with radical change. They suggest that the major difficulties associated with radical change is that the leaders, once they have determined what needed to be changed, do not develop an explicit strategy to lead the transition. These authors have developed a model to help leaders develop this transition strategy, and what they call the meta-strategy. They emphasize among others the inability of members of the organization to change their own frame of reference and indicate ways of facilitating the trimming.

They focus on the symbolic management and identification of actions that can be undertaken without causing a paralyzing resistance, which requires a very good knowledge of the organization, especially its culture and good political skills. It was about Tichy (1983) who in his famous book present a management model Strategic Change that made three interrelated components: cultural management, technical management, and policy management.

Cultural management (or symbolic) aims to influence the direction of change, including an evocative vision, clear and reliable communication, relevant links with the past, and recognition of success. Technical management can give a rational and practical foundation for change by planning rigorously, by establishing and realistic reading the structural changes and required systems, putting in place controls and monitoring. Finally, policy management is an often overlooked element, but it is especially critical in achieving a radical change. It requires knowledge and encourages its allies while neutralizing his opponents, and introduces positive change in individuals at key positions. These models, which have their uses, suggest one way to manage change and only give general guidance on the process to get there. Another approach suggests instead that there would be ways to manage radical change. Vandangeon-Derumez (1998) as a result of in-depth study of different cases where we conducted a radical change suggests that there are two patterns of changes: the directional change or prescribed and participatory or built change, each having its own logic.

The policy change is the traditional form of change where leaders define the direction to follow (that is to say what will change) and guide the implementation, which is the responsibility of middle managers and employees. By cons, in the case of participatory change management puts in place a process of change aimed at promoting the participation of members of the organization as to the definition of change (that is to say, the direction to follow) at its completion, the two often being simultaneous. The dynamic model of the cutting change in three phases: maturation, uprooting, and rooting, which take place in different ways depending on whether it is in the logical direction or participatory. The advantage of this model for the practice is to establish a series of activities to each step, which can be distinguished based on two modes of changes discussed earlier. The maturation phase is the step of preparing to change, and it includes the five following activities:

The stimulus identification let's see if the objective of the change is to seize an opportunity or respond to a problem.

(1) The search for information is essential to clarify the proposed change. The information can be collected by a prospective study or an internal audit. While the first is to anticipate the future, the second per- rather put a diagnosis on the situation of the company. It is based on these two activities we find the arguments when it comes time to justify the proposed change. These are controlled by the leader at the top, and are not associated with a particular mode change. First, the same change can be defined both as an opportunity to seize and as a way to solve a problem. Then a change, it is defined as a response to an opportunity or a threat, may cause a change directive or participatory approach.

(2) Awareness of the idea of change can be made through an announcement from the executive to inform members of the organization that change is considered. Sometimes, it is the announcement of the replacement of the management team that informs the rest of the organization that change is expected. Conversely, leaders can announce the start of a process of reflection on the change to which different members of the organization are invited to participate.

(3) In the case of an advertisement, the setting in motion of the organization is controlled by leaders who define the direction to be given to change, based on the vision they have developed. In the case of the start of a process of reflection, the members of the organization have the opportunity to participate in the definition of change, and the focus is then on the approach to adopt. This is when we identify the natural leaders who want to play an active role in the change process.

(4) Finally, the maturation phase may conclude with the finalization of a formal project or a process by which the project will be built collectively.

Thus, in a policy change, it is stimulated by an opportunity or a problem, awareness is through an ad, the actuation is controlled by the leaders, and focused on their vision that is articulated in a more formal project. Moreover, in a participative change, it involves people in a process where the focus is on the process, since the project is still unclear and remains unclear. The uprooting phase begins with the release of draft change throughout the organization and understands its implementation. It is built around the four following activities:

(1) The communication of the change project, when it comes to a formal project planned by management, is through a formal announcement that focuses on the message to make sure it is well understood. Furthermore, if a collective approach is business, communication is used to mobilize members of the organization to get involved in the change.

(2) We are witnessing, then, the implementation itself. In the case of a change orchestrated by the management, it can be brutal, as if the approach is interactive, the implementation is done gradually.

(3) The implementation of the change, even if it is planned and controlled, always leads to the development of local initiatives. In the case of a policy change, these initiatives come mostly from managers, are very close to the original plan, and closely supervised. Conversely, the participatory march aims to stimulate the generation of new initiatives, both by managers as employees, without much force in the creative process.

(4) Finally, the monitoring of the implementation is ensured by the hierarchy and accompanied by tools based training (TQM, technology, etc.), in the case of a change orchestrated by management. Moreover, in the collective process, not to impose change from above, monitoring is done by the natural leaders from different hierarchical levels, focuses on temporary structures (focus groups, round tables, etc.), and is accompanied by a broad training for comprehensive skills.

In summary, in a policy project, communicating the change through an official announcement, and implementation, whether sudden or gradual, above-mentioned converging initiatives developed by the hierarchy and is accompanied by a rather focused training on tools. Instead, participatory change promotes interaction, new ideas initiated by the base, training focused on reflection, and structural support.

Finally, the institutionalization phase of change, or rooting, includes the following three activities: (1) Evaluation of the actions, that is to say the achievement of a balance sheet of what has been undertaken now is the first activity. The leaders control evaluation in the case of a directive approach, whereas an interactive assessment, to collect the various points of view, will be implemented in a participatory process. (2) Following this review, it makes the necessary corrections to the coherence of activities redirecting the change to the original objectives, in the case of change orchestrated by leaders or by specifying and adjusting actions to avoid the dispersion in the case of the participatory change. (3) Stabilization of change is characterized, in the case of directional change, for implementing a stable working environment when the implementation is completed or, in the case of a participatory change, for clarification and formalization of the vision that developed in reflection and collective action.

Thus, in the process directive, in line with previous actions, the evaluation is done unidirectional by senior management, which is safe as that change is completed according to the original objectives, reviving, if necessary, a change that is winded by redirecting actions beyond the established framework then formalize the framework. Moreover, in the participatory model, the interactive assessment can lead senior management to treatment and adjustments to stabilize, in formalizing the vision that developed in action.

This model enables managers to consider how to implement these two different modes of change management: the logic directive where formulation and implementation of the change project are separated between policymakers and implementers, and participatory logic where leaders focus on collective action that enables the design and implementation of coexistence. However, the work of Derumez (1998) shows that many changes are in fact hybrids, the process of change is marked by the passage of a logic to another. This hybridization can be explained by the desire to avoid the limitations of each of these approaches. Thus, in a sense directive, the main problem lies in the transition from the corporate level to the operational level where the shareholders on the ground, during the implementation can adopt practices that are far from what was expected by management, and thus undermine the strategic coherence. The main problem of the participatory model is to create a coherent strategy from initiatives coming from everywhere in the organization. Leaders must then make choices that may demotivate members of the organization that no longer find themselves in the version finally adopted change.

The use of either of these two approaches is not determined by the type of transformation (shift, recovery, etc.). It is clear that the recovery intuitively seems more likely to be done according to the directive mode, as seen in the case of Iacocca Chrysler. However, the examples of Forges the Sorel, where leaders have involved the union in this business transformation approach the edge of bankruptcy, reflecting the fact that each case is unique. Moreover, among the cases listed by the author of the model, organizations in crisis do not always adopt the directive mode and those that do not necessarily favored proactive participation. However, this model provides clues to sensitize leaders to the particularities of context features that facilitate or hinder the conduct of operations in both modes of change management. It is therefore not here to favor one approach over another, but to be aware of the characteristics, advantages, and limitations of each.

Such a perspective on radical change in management does justice to the diversity of situations and organizational realities. It provides flexible tools for thinking about change in all its richness and complexity. Throughout this book, we are interested in strategic management. Traditionally, the business strategist is exercised by the executive or by a team of leaders. Following external and dull diagnoses, the leader establishes consistency between the elements it needs, willing or able to take into account. Once formulated strategy, it is communicated to other levels of the organization, and implementation tools, primarily a structure and appropriate management processes, facilitate its implementation. This process is often presented in a linear manner from formulation to implementation. It is also presented as one-time, localized in time, and lying before the action.

Despite some problems, among others, the unpredictability of the future and limited cognitive abilities of leaders, this process is possible in smaller organizations operating in relatively stable industries. But when the size of the company increases, the organization becomes complex, as in the case of diversified businesses, or that the environment becomes more complex, as in the case of global companies, the leader often needs help to exercise its correct strategist business. It is in this context very well described by Ansoff (1965), what appeared to be strategic planning. It also is in this context that since Porter systematic approaches such as the industrial economy have become increasingly important strategies. All of these tools to establish internal and external diagnostics business are useful to the leader and his management team. This is why the traditional approach to business strategists is still widespread, both in organizations and in business schools. This is what also explains the importance we have given to the planning of activities throughout the book, especially in Part II, "Designing the strategy." This part of the book focuses on how leaders can conduct an orderly strategic thinking, and discusses several tools that can help them achieve environmental analysis and organization, and to make appropriate policy choices.

An orderly strategic thinking activity, analyze and make strategic choices consistent with the elements of the analysis, plan activities, and programs based on these choices: this systematic procedure has many advantages. It leads the company to discipline you to think of the future; it forces you to have goals to reach and organize action to achieve these objectives; it provides the basis for evaluating the performance of organizational units and individuals; it creates order in the organization and is safe for executives, managers, and employees. But there are also problems with this approach; problems were experienced by several companies that have been identified by several authors (Morgan, 1983). A major problem lies in the fact that this process can disempower the leader leaving too much room to analysts. However, there are other problems that seem more important.

First, an approach to training strategies that implies that the company's leaders fail of the strategic competence of other members of the organization and the contribution they can make. On the other hand, an approach that wants a formalized process taking place mainly before the action may not be sensitive to strategies that may emerge during the action. It is then necessary to design the exercise of the strategist art can take another form. It is no longer for the leader to first formulate a strategy and then implement it. It is rather to facilitate the strategic action of all members of the organization and to establish an environment that enables their contribution to training strategies. It is more a one-off activity, but an action that is done in small steps, "along the way" in the words of Avenier (1997). And it is this process of this strategic action that daily strategy emerges.

Such an approach does not eliminate the role of leaders, on the contrary. As we have seen, the leaders remain the architects of the rationale and context of creators, and these roles are both important and demanding. It did not evacuate the importance of intentional strategies. These can be an element of the strategic action around what action of players develops; they can no longer be regarded as the business strategy. This approach to training strategies is particularly relevant in situations of complexity, turbulence and instability, when it is difficult to understand the context in which the organization operates, and the skills to cope. Some approaches attempt to reduce the complexity of the environment. This is what Porter chose to focus primarily on the competitive economic environment and reducing strategic diversity to three generic strategies. Other approaches recognize the complexity, but put forward simple rules that companies must adopt if they want to deal with it.

In both approaches we have just mentioned, and it was the same with the tools we have presented on the strategic management of diversified business and global business, it is still mainly the leaders who are responsible to find solutions that allow the company to operate and be successful in complex situations. In addressing the training strategies through strategic action of all members of the organization, we offer a different path. The strategy is a social construct, involving all members of the organization. They have a strategic competence flowing of learning they have acquired and tacit and explicit knowledge they have acquired over time. This knowledge is then put to use on a daily basis to solve problems that occur or for projects of all kinds. This is the daily work of all shareholders that emerges the organization's strategy.

Avenier (1997) prefers to talk about groping strategy. This strategy differs from incremental strategy, since it may allow radical changes. It is also different from the emerging strategy because it allows the realization of deliberate actions in emergent situations; it also promotes the emergence of deliberation, that is to say the emergence of deliberate projects. The tentative strategy is therefore built step by step, through multiple oscillations between reflection and action in a permanent tension between deliberate and emergent. It allows the organization to adapt and to adjust continuously in a changing environment. As we mentioned earlier, the leaders continue to play important roles, but they no longer act as the only designers of the strategy. As architects of the reason, they ensure that the shared representations correspond to the values and major objectives of the organization; they also ensure that the procedures and routines are consistent with this system of representations. As creators of context, they set up a context for participation and innovation, and they ensure that the structure and different management systems make participation and creativity possible.

It is through their leadership that leaders create an environment that enables members of the organization to participate actively and creatively, strategic action now. It is also thanks to their leadership that all members of the organization have to work together in the same direction, so the company is performing economically and socially. Designed in this way, the strategy formation process becomes a sense of inclusive process of building shared by the different levels of the organization, and inclusive management process in itself. The analysis and action become interrelated processes; the action can no longer be conceived as a simple implementation of strategic choices made at the top of the organization.

CHAPTER 12

Structures and Processes of Applied Management

At the turn of the century, Bombardier faces wrenching and crucial decisions for its future balance. Should it produce mid-size airplanes, and thus compete with the giants Airbus and Boeing? Would it not be better to strengthen its position in the small aircraft? Meanwhile, the main competitor of the company seems more determined than ever to dominate the field of activity of small planes and in doing so to give more power to go higher. And the giants give signals implying that they will fight without weakening against any intrusion on their chosen territory. Assuming that the company decides that its presence in airplanes' average sizes is necessary, the most important issues are related to how to achieve this and they involve multiple operational choices. So, should we ally with the new players to enter the sector, such as Chinese companies? Another interesting solution is to join a major player. But is it possible? In this case, what should we give?

Similarly, Alcan was recently found in front of a multitude of choices. First, since the emergence of China and India on the global economic scene, metal experts indicated that the future was favorable. It was necessary to strengthen the core business. Then, it was important to know how. Many solutions have emerged. Alcan could be integrated vertically in developing, either internally or by acquisition, the downstream or upstream of the manufacture of basic aluminum. It could also strengthen its competitiveness by increasing economies of scale through the acquisition of competitors. It could also diversify by considering other metals or other regions. The company could finally combine all these possibilities.

Alcan has chosen to remain in the field of aluminum and proceed with the acquisition of quality competitors such as the French firm Pechiney. Its greatest rivals, Alcoa, also adopted a similar decision. Then suddenly another actor who had chosen to work in several major metals, Rio Tinto, found in a much more favorable financial position and presented for the purchase of Alcan, which seemed too specialized to succeed in a world where all metals would become crucial. Investors have thus accepted the acquisition of Alcan by Rio Tinto and the disappearance of the largest Canadian business for over a century.

The operational choices are important decisions usually taken by the main shareholders of the company, sometimes individually, but more often collectively. These choices determine the competitiveness of the company and its survival. The operational choices do not always make changes as drastic as those mentioned for Bombardier and Alcan, but they are still crucial. We try to provide evidence to guide decision making. We begin by describing the difficulties of the decision process, and then we will focus on four patterns of operational choices, observed in several companies. We conclude with a brief discussion about how to evaluate the performance on these choices.

Manager is deciding. This is the decision that gives coherence and consistency to the organization. This will lead him/her to be interested in the anatomy of the decision, namely the distribution of decision-making functions in the organization between the rungs superiors (policymaking) and lower levels (business decisions). It will also look at the physiology of the decision, namely the different phases in the decision process. Simon's decision-making models distinguish three phases: (1) phase of intelligence or problem identification; (2) modeling phase or solutions designs; and (3) phases of choice or selection of the best solution, following the application of a "selection criterion."

It was taken for granted the existence of sufficient information on the range of options and their consequences, and the order of preferences. Several studies that have followed, including those of Simon himself, questioned the canonical model of decision making. First, we consider that the decision process develops within a system of constraints related to cognitive abilities, always limited, policymakers. This idea is at the heart of the bounded rationality model, the model of decision behavior in the company and the model of the dustbin (garbage can model).

Moreover, it states that the decision process is not based on a clear and consistent preference order or resulting in shared collective prefer- ences. Preferences are ambiguous, contradictory, and are not consistent. In addition, each reflected preferences in individual goals, according to its particular situation. So there exists conflicting individual goals and competing for the resources, and each seeks to uphold its own objectives and those of others to maintain their relative influence. It is in this context of bounded rationality of shareholders and political games that decisions must be taken. For some people, these constraints are so important that they prevent any rational decision based on a proper diagnosis of the context of business and organization. For others, these constraints lead to interest, not to deliberate decisions elaborated before action by leaders at the summit, but the decisions that emerge during the action and take account of a plurality of shareholders.

And for others, well, they require certain realism. It is in this current that we stand. They require certain realism. It is in this current that we stand. They require certain realism. It is in this current that we stand.

The organization is a place where all kinds of decisions are made. Sometimes, the leaders at the top are the only ones to make important operational decisions for the organization. This is often the case of entrepreneurs, when the company is small and young. But the decision by a single actor becomes more difficult as the size and complexity of the organization increases. The operational choices so often result from a complex and multifaceted interaction between top leaders, operational managers, and the board. By observing the behavior of companies, we can say that, of all the decisions taken by the leaders, foot types of decisions are of vital importance for the future of the organization: (1) decisions on mission organization: it is the institutional action; (2) decisions on areas in which the company wants to work: it is the action director; (3) decisions about how we hear compete with others in each of the areas of business activity: it is the business share; (4) decisions on the implementation mechanisms generated by previous decisions: these are the functional decisions.

The institutional action is the mission of the company. It includes the key objectives that the company wants to achieve and the values that drive its action. It corresponds to what the company wants to be and the image it wants to project to its stakeholders. This is kind of its personality and identity. This suggests that when the leaders of a company work on the formulation of an institutional decision, it will not change at the whim of leaders who will succeed the company's helm. This does not mean that, in times of major change in the direction of a company, the mission cannot be changed. After a crisis or a major turnaround, the company will want to let its customers and employees know that it is taking a new direction.

The institutional action is a guide and beacon at the time of formulation of the corporate action and business processes. Leaders should evaluate the decisions they are considering in the light of the principles and fundamental values contained in the mission of the company. In a sense, they must accept that their action is constrained by this mission. If the company's mission is never a reference point when making important operational decisions for the company, it is then only a cultural artifact, an inert object, a beautiful image you project to outside, and a public relations tool that has not much interest in optimizing management practice.

Here is an interesting case in this regard. The newspaper has a mission to be primarily a reflection of opinion and newspaper rather than a vehicle of information, and be active in the national political debate. The directors who have succeeded in the log header have all, in their own way; they have their action in respect of this mission. At one point, things have changed. The journal no longer give the mission to be a debate and opinion journal, but he defined himself as a witness, so as a newsletter mostly without political commitment. In doing so, the newspaper positioned itself in the same group as the existing large dailies.

The results are not immediate. While the situation of the newspaper had always been relatively weak compared to that of its competitors (due, among other things, and its low draft), it became catastrophic: the newspaper has experienced annual deficits that have increased year. The manager had to leave the management of the newspaper. He/she was replaced by a journalist who had worked in the company and who left it at the time of the shift. Returning to the original mission of the newspaper, while working to make it a more modern and efficient log, the CEO was able to restore the financial equilibrium of the company. This case is typical. It represents well the situation of all the great journalistic institutions. These media are living a big institutional change with the transformation of the property, which gradually passes into the hands of investors more concerned about the profitability by the values conveyed by the media. The journalistic elites are alarmed by what they see as damaging to society deviance. It is in a phase of great social changes, such as that experienced by the major capitalist countries, the pressure on mission statements is strongest.

Leaders constantly think about the future of their business. Do we want to increase the company's activities? Do we stay in the same areas of activity or do we want to move into new areas? Do we want to withdraw from certain areas? These are questions that pave the lives of business leaders. As we have seen, to answer these questions, managers may use a formal process of decision making, such as operational planning, or follow a less linear and less formal process. They can also help classificatory schemes whose heuristic value lies in their ability to name and classify reality, so their ability to bring order into complex realities.

There are three main guiding mechanisms: (1) the holding action; (2) growth action; and (3) the removal action. Each of these mechanisms may involve either the products or the markets. The holding action is a relative stability of the company's operations, both in terms of its products on the markets it serves. This decision is often considered relevant when the business is relatively stable, the company is doing well and resources are rather limited. But it could also be justified in all market situations. This is why a maintenance decision does not mean the status quo. Instead, a company that wants to maintain its activities in a market often has to work hard. In a growing market, maintenance may require a lot of investment.

In a more stable market, the company could be especially concerned about the renewal of its products. Thus, it seeks to improve the quality of its products and the attractiveness of its packaging. It could also try to find new functions in its products or to enter into agreements to market them under different brands. This is common in the food industry.

But there are situations, especially in the case of weakening of the market, where, to successfully protect its market share, the company will not only try to renew its products, but also perform a recovery of some or several of its ways. The holding action seems an easy decision to set up, since the makers are familiar with the products and markets in which the company is engaged. It is not so. Depending on the phase of the product life cycle and the company's position in its market, leaders will choose the same means of action. They will also have to decide how much human and financial resources will be needed to maintain the company's market share.

Thus, a maintenance decision can become very aggressive when the market declines. Indeed, when the company must maintain a very favorable position in a market that is fading, it has to make significant investments in research and development and marketing. This is the case of cigarette manufacturers, such as Imperial Tobacco. Due to anti-smoking advertising that presents the risks of smoking for the health of individuals, this market is in decline in North America. To maintain its very advantageous position, Imperial Tobacco has developed new types of cigarettes that says are less harmful to health; it also uses plenty of sponsorship of events and spends exorbitant amounts in the promotion.

A maintenance approach still presents some risks of "operational drowsiness." First, when the company managed to maintain its market share and that the investment rate of return is good, its leaders can be convinced that only incremental changes, and the margin is sufficient. On the other hand, leaders cannot be sufficiently attentive to profound changes in the context of business that may render inoperative in the future of a maintenance decision. Perhaps, one could argue that a company adopting a long holding action does not develop the skills and abilities necessary to compete in a complex and turbulent business context. A growth strategy is appropriate when the business context is favorable, the company is doing well and the leaders believe that the future of the company will yield growth. This appeal too many leaders. Such action can be achieved in several ways. Four ways for a company to grow: (1) the company can use its existing products to further penetrate the markets where it is already present (penetration); (2) it can use its current products to try to penetrate new markets (market expansion); (3) it can sell new products in existing markets (developing range); (4) finally, it may choose to diversify into new areas of activities, which are linked or not with its original field of activity.

An important operational decision for a company is to decide if it wants to grow in the field of activity where it already operates, or if it wants to move to other areas. A company that decides to grow by staying in its field can, by market penetration, market expansion or development of the product range. A company that wants to grow by choosing new areas takes a diversification decision. Companies decide to diversify to different reasons including the desire to spread the risk. Diversification can be connected or not connected. We say that diversification is connected when some of the skills acquired by the company in an area are transferable into new areas where it is committed. So, there is potential for synergy between the fields. Diversification is not connected when the new areas of activity require skills quite different from those that the company has, and little synergy can be established between them.

The matrix can be refined by speaking of current, new and expanded markets, and current products, new and improved. We find in the following table, a matrix of growth options and the list of actions to be taken according to the chosen operational options. It is possible to determine the benefits associated with different approaches "product" or "market." Some risks to growth are ahead. A growth approach is still very attractive for business. It has, moreover, as the holding action that we have spoken, the following difficulties:

- Market penetration is probably the action of growth that is easy to implement, since the product and the market are well-known leaders. But it's not always the case. In a mature market saturated, the decision can be difficult and expensive, since the acquisition of additional market share requires very high expenditure in terms of advertising and marketing. In a growing market, buying competitors may also require significant investments.
- Market expansion requires significant investment, but it requires above all knowledge about the state of competition in the new markets targeted and proven expertise in market analysis to fully understand the characteristics of new customers. Some successful companies have experienced great difficulties when they tried to settle out. For these companies, the market penetration became very expensive, since it already had a large market share. Market expansion, primarily local, then appeared as a desirable approach. The beginning was very difficult, laborious and still remains due to the difficulties of integrating acquisitions that were made.
- The expansion of the range by developing new products also requires major investments. Substantial amounts must be among others on research and development of new products, internally or externally, the purchase of licenses or patents to companies manufacturing these new products and the acquisition of firms. The accounting firms' successful enlargement of the range of products offered to customers. In addition to traditional audit activities, accounting firms have invested so much to offer various other services including management consulting and financing, outsourcing of various activities such as internal audit and payroll service. But the context may also evolve to challenge these actions. So, recently, due to the crisis sparked by the Enron scandal, many companies were forced to clearly separate their auditing activities from those related to the consultation. Sometimes, the action of growth may conflict with the position and the company's values. The case of the company "Birks" is a case in point. This company

specializes in the retail of jewelry and tableware, and had traditionally focused on high-end products. She tried to achieve growth by expanding its product range, adding mid-range products. This action proved to be a double failure because it also failed to attract new customers.

- Diversification is the riskiest growth mechanism. Indeed, it requires a lot of effort on the part of the company to get to understand the key success factors in new business areas where it wants to go. Moreover, it significantly increases the complexity of management. The development of structures required to complete the integration or simply good management of multiple activities that can be especially difficult. It is these difficulties that explain some spectacular failures experienced by several companies that have opted for a diversification decision.

Growth optimization is not easy to use. The case now Harlequin illustrates some of the challenges related to growth. Harlequin is changing in the industry of book publishing. Its sales and annual profits have been steadily increasing. In 1979, they were respectively $180 million and $20 million. Harlequin novels were published in 9 languages and sold in over 90 countries. Many observers then considered the company as the most profitable publisher in North America. As the romance novel market began to stagnate and Harlequin continued to have significant cash and had no debt, it had to invest for future growth. Harlequin decided to continue its geographic expansion in Japan, Scandinavia, Mexico, Venezuela, and Greece. It also decided to expand the range of its products by creating new collections for the North American public, German and Dutch. She also chose to diversify its activities by producing her first feature film, opening a store specializing in the sale of publications to teaching and buying a business selling by mail of toys, games, and small items of the kitchen.

However, despite the considerable financial resources that the company had to implement these decisions, these have not yielded the expected results or have failed altogether. By observing the operational behavior of this company, we have the impression that, because she had a lot of money, she felt obliged to use it, even when there was no opportunity to do so or that her skills were not clearly relevant for success in the new selected areas. Like what having much money to invest, following a very profitable growth can be a problem! The logic of growth is realized by the use of different operational maneuvers. Here are four operational tactics most commonly used by companies:

1. A business can achieve growth in its field with the acquisition of companies in the same field: Cascades growth is mainly due to its many paper manufacturer's acquisitions, the largest being that of Rolland Paper. A company can also carry out a merger with one or companies in the same field: this is what happened in the automotive industry with Daimler and Chrysler, Fiat and General Motors, and BMW and Rover. In all cases, it is horizontal integration. The mixed results of these consolidations show once again the difficulty of integrating different organizations.

2. A company can have an objective to control the downstream and upstream processes. This is a vertical integration decision. It can do this by developing itself in new business areas that will ensure the reliability of supplies or for the flow of its products. This is called internal development. It can also do this by acquiring companies that provide its inputs or marketing its outputs. Many companies make both internal development and acquisitions. Quebecor Company illustrates this situation. Initially a neighborhood newspaper publisher then was launched in the printing industry. It is further organically involved in the publication. Gradually, in order to reduce its dependence on suppliers and customers and continue to grow, it acquired other journals.

3. A company that aims to diversify the fact, usually, through the acquisition of companies active in the new sought-after areas, although organic developments (i.e. to say by internal development) are possible, as we mentioned in the example of Quebecor.

4. An organization may consider growth by a different route or that of alliances. Indeed, for its development, a firm can obtain the skills through alliances with firms that have them. These skills are of various kinds: technological, from research and development, procurement, distribution, customer service, etc. The Peugeot car manufacturer, because of its small size and its fear of disappearing in a merger with another manufacturer, continues to opt for an alliance decision. Thus, Peugeot cooperates with Fiat for more than 30 years and with Ford for 10 years to strengthen its global position in the manufacture of diesel engines.

The withdrawal approach is to when it means reducing or even stopping the activities of a company. This decision is relevant when the business context is unfavorable, the business goes wrong, and better opportunities exist in other areas. The company then decided to abandon certain products or activities or to withdraw partially or fully from certain markets. Life cycle phase in which the product or area of business activity and the relative position of the company in the market play a decisive role in decisions to revoke. Indeed, companies are adopting a withdrawal decision mainly when they do not manage to have a comfortable market position.

Like the steps of maintaining and growing the removal action may seem like an easy action to take. However, it is not. As with other approaches, it assumes to know the life cycle of the business areas in which the company is engaged as well as the evolution of competition in these areas. But the particular risk associated with the removal action is of another order. On the one hand, by withdrawing from certain activities, the company made to rest some fixed costs over fewer activities. Moreover, the withdrawal of actions can have negative financial impacts that are greater than the savings achieved by the removal of certain activities. For example, a withdrawal decision may harm the image of the company. This will have a negative impact on other activities. Finally, a withdrawal decision can be demotivating for staff and weaken the company's portfolio of skills, especially if the decision is accompanied by the departure of employees and key executives. The withdrawal can take three main forms: the entrenchment of activities, sale of businesses, and liquidation.

The entrenchment of eliminating certain products or services. This was the case when a railway company decided to eliminate a number of destinations it offered before. This was also the case when the company Eaton has cut its retail activities catalog sales and subsequently, sales of household appliances. This is also the case of the Montreal Exchange, which has decided to sever its activity equities and bonds to focus on the options market. The products and services that are usually subtracted deemed unprofitable activities for the company, or are less profitable than other activities in which it is involved. The more competition there is (in the industry), the less the company can maintain (little or no) profitable products and services.

This is true not only for companies that pursue an action of cost leadership, but also for those adopting a decision of differentiation. The sale of activities is for the company to withdraw completely or partially from an area in which it was active before. In this sense, it is a form of entrenchment. The company uses this operational maneuver in two circumstances: first, when it believes it will be able to strengthen its position in a field of activity, given the competition that exists and instead it has managed to occupy in the market; second, when it wants to free up resources, mainly financial.

After the wave of diversification, several companies, financial results disappointed, decided to focus their activities in their areas of origin. They therefore massively divested. This is the case of "Canam Group." Originally, the group was active in the steel. Very quickly, it has diversified into a complementary domain to steel, that of the semitrailers. Then, the group has diversified into the field of office furniture. The financial results of the company were excellent. Under the effect of the mode of diversification, Canam has decided to take an important position in Noverco, a company that was the architect of the investment policy and diversification of GazM. This significant investment forced the—Canam Group—to divest areas semitrailers and office furniture. Realizing that achieving effective control of Noverco force it to borrow significantly, Marcel Dutil, leader of Canam Group, has chosen to divest the energy sector.

These activities of sales movements are an important part of operational maneuvers in which companies now have. The liquidation is to divest itself completely of a company. The company adopts the rare decision when forced to do so. This is the case when the company is no longer viable and is experiencing significant financial difficulties. This is certainly the case when financial difficulties are such that they are forcing the company into bankruptcy. That's what happened to Eaton. The recovery action that the company tried to set up did not work and it found itself obliged to liquidate the company.

Complex enterprises are usually present in several areas. Their manager decision is therefore a combined decision, that is to say a decision that combines decisions continued in different areas of business activity. A company may pursue a maintenance decision in areas that are mature and in which it has managed to carve a comfortable place; it can also pursue a decision or termination if it considers that some of its fields have become less interesting or as other areas seem promising it; it can finally pursue a growth making in the activities it deems promising. When a company is engaged in several areas, portfolio analysis is a valuable tool for decision support.

The popularity of private companies for portfolio analysis models can be explained by the ability of these to simplify a reality more and more complex and thus to help the leaders of the diversified businesses make operational choices that are imposed. Leaders rarely possess a thorough knowledge of all areas of activity, varied and numerous, in which the business is located. They must also decide the future of each of these areas and add new areas of activity. The portfolio analysis models then help them clarify their criteria.

The relative market share is measured by the ratio of the company's market share compared to the market share of the nearest competitor. This ratio reflects the competitive position of the company regarding the costs, benefits in terms of costs arising from a larger volume of business than that of competitors. The quantitative model is a purely financial logic. The optimal allocation of the company's financial resources is done taking into account the profitability of different business areas and their liquidity needs. In a diversified company managed as a portfolio, business decisions are developed from each of the business areas, while the corporate action is developed from the head office.

That's where leaders use the portfolio model to make operational choices required regarding maintenance, development or abandoning their various areas of activity, as well as the addition of new areas. The leaders of all interested in the return on capital invested in the various fields of activity, and not the possible synergy between these areas. Unlike quantitative models in which quality does not rely on the only financial sense, it takes into account qualitative factors. It is interested in the market but also the company as a whole. The leaders of all interested in the return on capital invested in the various fields of activity, and not the possible synergy between these areas.

The first axis of the model is the value of a field of activity for the company, including not only the domain of the growth rate of activity, but also the interest in that area for the company. Can there be synergy between this field of activity and other business areas? How now to control its already key factors for success in this area? This area of activity helps later acquire a transferable experience to other areas? The value of a field of activity is also related to the company's capabilities and its identity: its distinctive competencies, expertise, interests, and its "will to do." The second axis is the competitive position of the company in a business area, its position in the life cycle, and the risks it entails for the company. It is therefore no more than limit the analysis to the only relative market share.

The search for consistency, although this model is more qualitative than the previous one, it can be very useful for leaders of diversified companies to make operational choices. It allows them to make these choices only based not only on financial flows of the company, but also on all the features that help forge the identity of a company and its ability to compete with other companies. For each of the business areas in which it is engaged, the company must decide how it intends to fight against competitors or, in other words, how it intends to position itself in relation to these. The company has two major decisions regarding its positioning.

First, it must discover and develop a way to compete with other companies that is unique and different from those of its competitors, and that is some continuity. Porter notes two ways to obtain a sustainable competitive advantage: cost leadership and differentiation. On the other hand, the company must choose its operational target, that is to say the extent of the market it is in a given industry. This target can be broad (wide variety of products and services to all customer segments) or narrow.

These two dimensions, the type of competitive advantage and scope of the operational target, define what Porter calls the generic decisions, namely three fundamentally different ways to position and compete with other companies in a given industry: the action of cost leadership, the action of differentiation, and concentration of action, on a close target, may use the cost leadership or differentiation. Companies can be successful in adopting different types of positioning in an industry. We can say that there is no one best way, that is to say one right way to compete with other companies in a given industrial sort. It is well away from some of its predecessors who had the experience curve and economies of scale only profitable positioning on an industry basis.

The cost leadership is to build an organization that can have lower costs than its competitors. It is to design, produce and market a product or a service comparable to that of its competitors, but at lower costs. This allows the seller the product or service at lower prices than its competitors, the same price that the latter, releasing a higher margin. This is possible due to economies of scale and scope that arise from volume production and through tight control of fixed and variable costs. This approach may require, in some areas, a lot of capital and significant engineering resources to simplify the design of ducts pro and make them less expensive to produce, automate production processes, and to extend the distribution channels. All companies must control costs in order to identify the highest margin possible: re-engineering or just-in-time manufacturing are approaches and techniques that can help a company to redefine its process of production and marketing to make it more efficient and less expensive. They help to improve the operational efficiency of the company. But this does not mean that these companies continue dominance of orientation costs. While cost control is oriented action internally, the cost leadership is primarily a positioning action externally.

All companies must control costs in order to identify the highest margin possible: re-engineering or just-in-time manufacturing are approaches and techniques that can help a company to redefine its process of production and marketing to the make it more efficient and less expensive. They help to improve the operational efficiency of the company. But this does not mean that these companies continue dominance of orientation costs. While cost control is oriented action internally, the cost leadership is primarily a positioning action externally. All companies must control costs in order to identify the highest margin possible: re-engineering or just-in-time manufacturing are approaches and techniques that can help a company to redefine its process of production and marketing the make it more efficient and less expensive. They help to improve the operational efficiency of the company. But this does not mean that these companies continue dominance of orientation costs.

While cost control is oriented action internally, the cost leadership is primarily a positioning action externally. Re-engineering or just-in-time manufacturing are approaches and techniques that can help a company to redefine its process of production and marketing in order to make it more efficient and less expensive. They help to improve the operational efficiency of the company. But this does not mean that these companies continue dominance of orientation costs. While cost control is oriented action internally, the cost leadership is primarily a positioning action externally.

The action of the company multi-Marques is a decision of cost leadership. Multi-Marques are moving in the baking industry. This is a highly concentrated industry, with few suppliers, a small number of competitors who control almost all of the market and customers are mainly supermarkets. In this industry, Multi-Marques had set a goal to be number one. To achieve this, the company was always aware that she would pursue a decision of cost leadership, backed up by production volumes. After 50 mergers and acquisitions, Multi-Marques acquired a volume that allows it to reduce its costs significantly. By centralizing the head office, among others, procurement decisions, negotiations with customers and choices relating to the product range, the company managed to have much lower costs than those of its competitors and tablets most interesting areas in the stores. This decision enabled multi-Marques to become a very profitable business and the largest bakery in the region, ahead of its main Canadian competitor, Weston.

A decision differentiation is to produce a good or a service that has a unique character for the customer, for which it is willing to pay more. The uniqueness of a product or a service may result from the type of materials used and their quality, product design, performance or image it projects. It can also result from the distribution network used, after-sales service or warranty that is attached to the product. The customer is sometimes willing to pay more for a particular product or service as long as it perceives that the added value is greater than the price it would pay for a product that does not have such features.

Differentiation action usually requires a lot of resources for the creation and development of products to ensure quality features, performance or larger than those of competing products reliability. Many resources have to be devoted to marketing to create a brand for those products. The action taken by Hermes is a decision of differentiation. Hermes is a luxury business. It started in leather goods, especially in manufacturing "saddles." It now affects 12 businesses, the main ones being leather goods, silk, ready-to-wear, perfumes, watches, the tableware, and jewelry. In all cases, Hermès products are distinguished by the wide quality of raw materials and design, a handcrafted, through advertising and highly selective distribution channels and a very strong brand. In other words, all the elements of the value chain are consistent with the pursuit of prestige and exclusivity.

The Hermes differentiation action is accompanied by high prices, customers agree to pay. This is a focused differentiation decision on a narrow segment of the market, that is, customers seeking the high end. The profitability of Hermes is exceptional, and that profitability will continue as long as consumers will feel that the value of the products justifies the high prices charged by the company. Companies such as Rolls-Royce, Mercedes, and BMW are other examples of focusing by differentiation. Unlike Hermes, the company Cray Research has not been able to maintain, profitably, the action of differentiation it had adopted at the time of the creation of the company.

The latter, a computer genius, stroking a big dream: build the most powerful computer in the world. The excellent reputation of the founder of Cray Research has allowed the company, even in the absence of a tangible product to benefit from significant financial resources. After designing supercomputers Cray 1 and 2, it has addressed the design of Cray 3. The action of the company was, clearly, in a concentration of decision (it was interested primarily in the laboratory research) by differentiation (building fast supercomputers). Cray is very concerned about the needs of other types of clients, such as companies looking for not only fast computers but could also be used in a friendly way. This was not the case at the Cray 2, already in operation, or the Cray 3, in development.

These industrial customers, who represented a growing market, were no longer willing to pay for a product with a very high price that was not justified by the value that the product had for them. They gradually abandoned Cray and its technological dreams of greatness for more sensitive to their manufacturers' needs, forcing the departure of Seymour Cray, and repositioning of the company Cray Research. This measure of cost leadership and differentiation meet two logics: that of the low cost and the high prices. A company can gain a competitive advantage by having lower costs than those of its competitors; its competitive advantage may also result from its ability to charge higher prices for its products and services. But can a single company simultaneously pursue both actions? To this question, we must respond in the negative. Indeed, designing, producing, and thinking a unique product that has significant added value for the consumer can be achieved with additional costs.

Conversely, the cost of the plan on the leader usually requires that the company agrees not to offer guests all the additions that exceed the desired functionality. The company must make a clear choice in favor of one or the other of these decisions. This was the problem of a company like Eaton, who prosecuted both a decision specialized stores offering premium products and a big box store decision offering cheap products.

However, pursuing a cost by the domination of decision does not mean ignoring what could help to distinguish itself from its competitors. Similarly, a decision to pursue differentiation does not mean we can ignore everything that could reduce its costs. According to Porter, any optimization must be concerned with both the relative costs and relative differentiation.

Champions' low costs should have an acceptable quality of the product, and the champions of differentiation should closely monitor the costs of activities that are not directly related to the customer value chain. It is therefore possible to make progress on both fronts simultaneously. It's not just choosing a location decision, it is still necessary that the company will stick to its choice of business decision. When Porter analyzes very successful companies, such as Walmart, Sony or Crown Cork and Seal, it concludes that what characterizes them is that they are very consistent in their core business choices and they try constantly to improve the way to put them into practice.

Companies that are successful are able to take advantage of advances in technology, to innovate and improve in order to reduce costs or to better differentiate relative to their competitors. But they pursue relentlessly the way of choices they have made. Is that they are very consistent in their core business choices and they are trying to constantly improve the way to put them into practice?

Changing business decisions is always very difficult. Changing its image among distribution channels and with consumers is quite difficult, but the company can still do it by operating under different brands. Moreover, what is more difficult is to change the organization. An organization built through time, skills to implement a decision of cost leadership or differentiation. The skills specific to each of these decisions are radically different. Changing business decision therefore means renouncing its learned skills and developing new skills, which is, in the opinion of all, a real challenge. The three types of decisions, one must add the functional mechanisms. Functional decisions are attached to the main functions of the company, namely the marketing, production, human resources, finance, research and development, procurement, monitoring, and management information.

Obviously functional decisions are especially important when implementing the guiding mechanisms and business activities, since these are often functional decisions that enable the successful implementation of the latter. This is why the leaders of a company should be concerned with functional mechanisms at a time when they make business decisions and directors. They thus ensure they are consistent with the choices they make, they can support these choices or, if this is not the case.

The director chosen action has an effect on the functional decisions. In the field of the decision, literature attaches special importance to competitive decisions. The operational choices (both institutional decisions, directors, business, and functional) are aimed at enabling the company to "beat" the competition. According to Porter, for example, it is so important to maintain competition that it opposes all that is likely to decrease the intensity. As we have seen, the context of business of the company is becoming more complex and differentiated: it consists of several players for following different mechanisms, while being increasingly interdependent situations, the ones with the others. The business context is also increasingly turbulent and unpredictable: it is changing fast and often difficult to predict. The company is then faced with economic uncertainties, technological, and political.

It is in this context of uncertainty that businesses come to develop cooperation mechanisms in order to be better able to face the competition. One speaks in this case, exchange relationships between organizations, collaboration agreements, hybrid arrangements, collective decisions, and operational covenants. The decision for a company to engage in an operational alliance of flow analyzes that leaders are the opportunities and threats in the context of the business case and internal capacities of the company to counter threats and take advantage of opportunities. Operational alliances can therefore be considered as a likely way to help the company improve its competitive position and realize its decisions or business manager. Alliances are generally for carrying out functional activities, such as R & D, and rarely for market penetration. Regulatory bodies shall ensure that such practices do not just generate the collusion on the backs of customers.

Company executives make operational choices, develop institutional mechanisms, Director, Business, and functional so that their business maintains or increases its performance. To be successful, the company must do well, referring to different financial criteria. But if thereof are requirements for performance, they are not sufficient. A business is successful if, in addition to good financial results it obtains, it is able to transform itself to cope with changes in context of business. As the operational management is the process by which managers ensure the long-term adaptation of the business-to-business context, the only really useful performance metrics are those that assess the adaptability of the firm. A well-suited company has a decision consistent with the structure and competitive dynamics of the industry; it has an organizational structure consistent with the context of dealing with the selected action; it consistent management systems with action and organizational structure.

Finally, it has an appropriate management style in the operational context in which the business is located. Ultimately, a well-adapted firm must be able to match forces with the context of the opportunities and align its various administrative systems with the action it has chosen. It must be said that the traditional performance measures based solely on the profitability of the firm, and are inadequate to assess the operating performance of the company. As performance is a complex phenomenon, so use several indicators to define. It focuses on two measures that discriminate operationally successful companies and those that are not.

The first of these measures for assessing the quality of transformations that occur in the business: on one side, it is to assess the company's ability to "exploit" so profitable context of the case and to ensure that the contributions of different stakeholders of the company exceed that it gives them rewards for their cooperation; on the other hand, it is evaluating corporate investments, from its surplus resources (slack resources) to improve its ability to adapt to an environment of uncertain future case and unknown.

The second measure used to measure the degree of satisfaction of all stakeholders of the company and not only those of shareholders. Here, we are plunged into the heart of the debate that continues to be used on the company's performance. Both in private and public sector performances, it has often been discussed in terms of effectiveness and efficiency. Evaluating the effectiveness of a company is for the leaders to question whether the company they lead doing the right things. Traditionally, effectiveness was evaluated based on expectations of shareholders alone. The question that arose was: do they meet the company's financial results and shareholder expectations? For leaders, to ask if the company they lead doing the right things. Traditionally, effectiveness was evaluated based on expectations of shareholders alone. The question that arose was: do they meet the company's financial results and shareholder expectations? For leaders, to ask if the company they lead doing the right things. Traditionally, effectiveness was evaluated based on expectations of shareholders alone. The question that arose was: do they meet the company's financial results and shareholder expectations?

The evaluation of effectiveness must now take into account the expectations of stakeholders. The question then is whether the financial and social results match the expectations of shareholders and those of other stakeholders. Our mission, as formulated, does it reflect the great goals we pursue and the values to support it? Is it consistent with the expectations of our shareholders and with other stakeholders? The business areas in which the company is involved, and the choices maintenance, growth or shrinkage that are made, do they correspond to the expectations of shareholders and those of other stakeholders? Do the business actions we are pursuing most likely to meet the expectations of our customers?

The question of the effectiveness inevitably leads executives to assess the impacts of the mechanisms adopted by their company. The criteria used to measure these impacts will vary, however, significantly, depending on whether they are interested in impacts for the shareholders or impacts on other stakeholders. Indeed, shareholders, whose compensation is based on investment, are mainly interested in the economic and financial indicators of the company and the share price.

In some cases, shareholder expectations therefore may differ from those of managers. While managers are representatives of the shareholders within the company, they must implement the orientations favored by the latter, they also have their own design mechanisms that decision Entre should adopt. They can therefore focus on growth making, market share growth that does not translate into an increase in profitability of the company. As other stakeholders, they are primarily interested in other performance criteria. Consider the following three types of stakeholders: customers, employees, and environmental groups. Customers evaluate the performance of the company based on products and services that it provides them, and in the light of their expectations regarding quality, cost, or book value. Their assessment of this performance translates into purchase behavior and loyalty by relating brands as for employees.

Finally, environmental groups are interested in the impact of the company's activities in terms of the context of business and sustainable development. The criteria they hold are not financial; rather they relate to the satisfaction of the citizens. For us, the operational management of the company cannot be considered appropriate if, to define the efficiency, it is focused exclusively on the interests of shareholders. The evaluation of the efficiency of the company is for the leaders to summon if the company does things. More specifically, the issue of efficiency leads them to question the functional business decisions: Do our production systems enable us to produce the best possible price? Is our distribution system adequate? Do we have the qualified personnel we need?

The issue of efficiency is inseparable from the analysis that leaders must make the capabilities and business skills. To assist in this effort, leaders may use competitive benchmarking (benchmarking), which allows a company to compare its performance with that of companies in the same industry, in particular the most successful companies. The evaluation of the performance of the company, in terms of effectiveness and efficiency, encourages managers to make operational choices that can make significant changes. The transformations can be performed in a more or less large manner depending on whether to change the beliefs and values of the company, its position, or its practices. Otherwise, action can also be a pattern emerging from the action. In this case, the operational choices, more than planned to be a priori, are in course of action. In such a context, structure, culture and leadership, which constitute the framework of operational activities greatly influence the choice. In fact, they are seen as levers to achieve the planned action or as the framework for action that emerges, structure, culture and leadership that are essential components of the optimization of business.

CHAPTER 13

Operational Values for Management Optimization

The organizational structure is both a lever to build a competitive advantage and an implementation tool. First, we define what we mean by organizational structure. Second, we will focus on the relationship operational structure approach. Third, we will deal with key management processes used by management to achieve operational approach. In a last time, we will discuss the structure as a framework for strategic action. The structure is the set of functions and determining relationships formally missions that each unit of the organization must accomplish, and modes of cooperation between these units. When the size of organizations than two or three people they endow with a structure that allows to divide the work by function, by product, or territory and to better coordinate and oversee the efforts of all. When they are structured, organizations must:

- Choose between differentiation and integration. There are two extremes: one could have a different position for each individual; we could also have a single type of position for all individuals. We know very well that to be effective, the structure of the organization should be between these two extremes. The questions that arise are the following: On what basis should we specialize (function, product/service, customers, region, etc.)? What activities are different enough to be separated? How far should you specialize? How to manage critical interdependencies between activities and products? What linkages should be put in place?
- Determining the degree of control and autonomy of the various elements of the structure. The structure should be neither too loose nor too tight. Every broad and complex enterprise faces a Fundamental paradox. On the one hand, senior managers need to be sure that, in a competitive and tough business context, they are

positioned to pull the levers that result in an adequate and timely response to key exchange. On the other hand, they must zealously guard against imposing controls so rigid as to choke the life from the organization.

Among the important variables, there are two that correspond to the two dilemmas we talked about earlier: the division of labor and coordination of work. In terms of division of labor, five elements describe all organizations: (1) the strategic apex (usually senior management and those who assist him directly); (2) operational core, composed of the people who produce the services or products that are the purpose of the organization; (3) techno structure or all professionals whose mission is to establish the standards (of work, results, expertise) for others; (4) support staff, which carries out activities that are not related to the primary mission of the organization and, ultimately, could be obtained from the outside; and (5) the hierarchy, which appears when the organization takes an important dimension. And in coordination, five modes are used by organizations: (1) direct supervision; (2) the mutual adjustment; (3) standardization work; (4) standardization of expertise; and (5) standardization of results.

The five parts of the organization and the five main modes of coordination combine "natural way" to give five generic structures:

1. The simple structure is a structure where the dominant part is the strategic apex, and where the main mode of coordination is direct supervision. Generally, this kind of structure is much formalized. There is no techno, no hierarchy, nor often support staff. This structure is very suitable for simple and rapid innovations in changing environments.
2. The red tape is a structure where the main mode of coordination is the standardization of work, and where the ruling party is techno. In this case, the organization is highly developed with substantial hierarchy and a large support staff. This kind of structure is suitable for mass production, in the case of stable contexts.
3. The professional bureaucracy is a structure where the ruling party is the operational core, usually formed of professionals, and where the main mode of coordination is the standardization of skills. There are usually no techno because professionals resist any attempt to standardize the work. However, the support staff tends

to be very important. This kind of structure is well suited to activities that require a complex know-how, in a context of business that is relatively stable.

4. The male dominance is a structure where the main mode of coordination is mutual adjustment, and where the dominant part is then the support staff, which is the permanent part of the organization. There is in this structure that few techno, and the organization is generally in constant flux, with temporary workers combinations for relaying of temporary needs. This kind of structure is very suitable to accomplish unique tasks and therefore innovation.

5. The divisional structure is one whose units may be structures of all other types. It is a form in which the main mode of coordination is the standardization of results and, consequently, where the ruling party is the hierarchy. This structure is very well adapted to situations where activities are multiple and diversified.

Some authors, related to the contingency theory, sought to establish links between the structure and certain environmental and organizational factors. It must be said that the structure of a firm is closely linked to its production technology system. Thus, mass production went well with a formal structure, while companies with a customized production or automated process tended to be organized in a more flexible manner. Also, the context of business plays an important role in determining the structure. For example, container companies, the environment was simple and stable, had a structure based on standardization and direct supervision. However, plastic companies who faced a more complex and dynamic business context had a structure based on coordination by mutual adjustment. The firms in the food sector, meanwhile, had a terraced structure.

Also all the mechanisms of the operation, including the design of the organization (meant by this strategic positioning), the structure, assessment mechanisms, and people management, to be co-aligned with the demands and uncertainties of the environment. This co-alignment has come to be known as the fit, which means compatibility, consistency, and fit including the design of the organization (meant by this strategic positioning), the structure, assessment mechanisms, and people management, to be co-aligned with the demands and uncertainties of the environment.

In this regard, what is the nature of the relationship between the operational approach and structure? The operational approach is directly related to what was happening in the environment, specifically in the market. More importantly, once the operational approach chosen, all the mechanisms of the operation, and in particular the structure, are forced by this operational approach. First, we describe the case of DuPont, which served as a basis to work from Chandler. Then we will show how Chandler's theory on the relationship between operational approach and structure has emerged in the area of operational approach.

DuPont was at the beginning of the 20th century, a company producing explosives, dynamite, and black powder in particular. It was led by cousins Bridge. Eugene, the president, was an entrepreneur, an "empire builder" in the words of Chandler. He acquired many small businesses and explosives factories, increasing considerably the size and scope of DuPont. Eugene was not interested in management. He neglected this aspect and was concerned only development. He ran each plant separately, by personally appointing directors in which he had a personal confidence. Thus, the company was run as a family, each plant manager taking care of everything locally, including the production and distribution sales in a particular territory, and being accountable to the President. There was no overall coordination.

From the birth of the company until 1990, we responded to market opportunities by building or buying explosive manufacturing plants. The operational approach was simple and was to grow as quickly as possible. The structural arrangements were dominated by a simple structure, with a chef and a host of collaborators. All management systems remain informal and uncoordinated. The operational approach was simple and was to grow as quickly as possible. The structural arrangements were dominated by a simple structure, with a chef and a host of collaborators. All management systems remain informal and uncoordinated. The operational approach was simple and was to grow as quickly as possible. The structural arrangements were dominated by a simple structure, with a chef and a host of collabora-tors. All management systems remain informal and uncoordinated.

As long as the competition was weak, this management was acceptable, the company generating the profits needed to continue its development. But gradually, serious competitors started to appear and they were large enough so that it is not possible to easily consider acquisition. DuPont then appeared as disabled because of his inability to coordinate its activities. On the death of Eugene, the new president, and especially Pierre du Pont, the treasurer, tried to quickly consolidate plant operations, closing some of them and building in other where it was more appropriate. But more importantly, it was coordinating the operation of this great together to reduce costs and optimize inventory and supplies. Peter then began to put in place a structure that better meet the operational approach to growth in all directions of his cousin. This resulted in a structure known today as the "centralized functional structure."

This new type of structure allows putting together the production operations and thus optimizes properly for sharing expertise and reducing costs. The same model was used for marketing. Markets were taken together, and supply was considered a general case rather than a shared responsibility regionally. Peter then began to put in place a structure that better meet the operational approach to growth in all directions of his cousin. This resulted in a structure known today as the "centralized functional structure."

Finally, the administration has been formalized: more information is gathered on costs, margins obtained, the rate of return on investment, etc. Performance measurement and compensation managers have become more systematic and linked to predetermined schedules. We asked each manager not to generate profit but to achieve functional goals which then would make a profit for the in-seems the company. Later, the president's office took over large to allow to centrally manage a company became much more professional in its practices. These structural arrangements were all geared toward centralizing the management and planning of the operation.

This was necessary due to the fact that the specialization of functions to the President left no responsibility for the whole. This new structure has served remarkably the company, which then had its best time, with both planned and exceptional growth unparalleled profitability. With its collaborators, the architect of this impressive rehabilitation, Pierre du Pont, now president, has refined the company's management mechanisms to the point that his achievements have become classics taught in business schools.

The company was so successful that production reached record highs. The outbreak of the First World War was the product when the company was at the zenith of his fame, controlling much of the US market and arousing the suspicion of antitrust authorities. The company has also been forced to split into several companies as a result of the application of the Sherman Act Antitrust Law.

The vast production of wartime, however, raised issues previously unknown. For example, by-products of explosives, once considered waste and sold or distributed freely that wanted them, were now in such large quantities that there was more than enough market or proper disposal system. The company then decided to examine the possibilities of using these by-products for commercial purposes.

These by-products were aromatic benzene-type, toluene, etc., the base today fine chemicals and plastics. The leaders then embarked on the manufacture of such products, including dyes, nylon, leather, and synthetic rubber, etc. Their resources, both human and financial, allowed to be optimistic all development paths. The company is solid and promising future, Pierre decided it was time for him to hand over the presidency to his brother Irenaeus and to go to other places. For DuPont, the problems would soon begin. By engaging in manufacturing new fine chemicals, the company had to change the field of activity, but it seemed not really realize it and continued to work with the old centralized functional structure problems would soon begin.

As new products were totally different, they would need radically different production processes. They were, moreover, in quantities much lower, for poorly understood markets. The markets fine chemicals and plastics were completely different markets of explosives markets. Customers of these were few, but were experts in handling and using the products. It was enough to produce, distribute, and provide minimal assistance in that related to the storage and handling of products so that customers are satisfied. In the case of new products, on the contrary, customers were very numerous, unsophisticated, knowing evil all possible uses and manipulation, and even product features. It would therefore be necessary to give their hand so they know how to use them.

As DuPont was dominated by functions, the leaders of the various functions necessarily accorded full attention to old products, the more lucrative, easier to sell, representing almost all of their income, and profits. So they ignored the new products and new markets. DuPont, then the company best managed and most admired, was losing money with all its new products.

The young leaders were well aware that there was a problem of structural arrangement, but the president did not want to hear about a structural change. Why, he said, should we change what has served us so well in the past? It took DuPont decides to consider a new structure.

This structure, the greatest innovations by Chandler, acknowledged the differences between the products and the need to manage separately. Each product line became a division with its own functions (production, sales, and administration) and was run as a separate company, but with a firm's overall coordination for financial matters and personnel management. This structure is called "decentralized divisional structure." With this new organization, the company would dominate the US market and the global market for chemicals, as she had previously dominated the explosives.

The story from DuPont is most revealing of the theory of Chandler, he stated as follows: "The operational approach above the structure." This means that, when adopting an operational approach, it is obliged to adapt the structure accordingly. Chandler developed his theory by studying four major companies: General Motors, Standard Oil of New Jersey (the predecessor of Exxon), Sears, Roebuck & Co. and, of course, DuPont. But he confirmed his results by studying statistically many other successful companies. Its results were also confirmed by Héau work in France, Channon UK, Germany and Thanheiser by those of his own research team in Japan. Others stressed the firm's development process that resulted from such a theory. So, Scott Salter then proposed an evolution of the company into three phases. Each stage combines a type of operational approach to a type of structure, but these authors also showed that all other management mechanisms were affected. These three phases with two sophistications called "conglomerate" and "global structure" can be considered a kind of business life cycle.

A fourth phase has also been proposed to account for situations where the functional aspects and aspects of markets must be combined. This phase introduces particular structure called "matrix." Each stage combines a type of operational approach to a type of structure, but these authors also showed that all other management mechanisms were affected. These three phases with two sophistications called "conglomerate" and "global structure" can be considered a kind of business life cycle. A fourth phase has also been proposed to account for situations where the functional aspects and aspects of markets must be combined.

The work that is most directly linked to that of Chandler, and that had a lot of echo, overall for diversification by mergers and acquisitions, is the Rumelt. This author has undertaken to demonstrate empirically the relationship between the operational approach and structure, and between the latter and performance. His work, although methodologically controversial shows that operational approach and structure are closely linked and that the quality of their association is a determinant of performance. It is interesting to return to the types of structures that we discussed earlier, particularly the functional structure and the divisional structure, and show what they are relevant strategic context.

The centralized functional structure has several advantages when a company operates in a single field of activity: it provides a clear definition of responsibilities; it comprises simple control mechanisms; the leader remains in contact with all operations; and is effective in terms of resources. It is a structure that allows the organization to be successful. This structure, however, poses a number of problems: it generates narrow logic of functional specialties; it stiffens the overall operation by raising the formalism and bureaucratic behavior. Leaders have great difficulty in integrating the specialized activities into a coherent and dynamic whole. They can also be overwhelmed by the operations and routine problems. To remain successful, a functional structure must develop mechanisms for consultation and dialogue, which will offset the disadvantages inherent in this type of structure. As for the decentralized divisional structure, it is appropriate when the company diversifies and works in several areas of activity or in multiple geographic regions.

The divisional structure, with the creation of divisions by area or region, has several advantages over the functional structure: it allows to decentralize the general management initiative; it facilitates the comparison of units; it helps develop skills related to the peculiarities of each division; it gives off the leaders at the top of the daily management concerns and allows them to focus their time and energy to strategic thinking. But this structure also poses a number of problems: it can lead to duplication of resources; it creates difficult market segments logical to reconcile, especially in terms of resource allocation.

The company therefore has no choice but to develop liaison mechanisms between its divisions in order to create synergies and facilitate reciprocal learning. General Electric, under the direction of Welch, seems to have realized this learning admirably. However, the rule derived Chandler's work is always true regardless of the type of structure adopted.

Regarding the management process, they are important for the achievement of operational approaches. The leaders therefore must wear their special attention. It is by adjusting a number of management tools that these processes take shape and become consistent with the objectives chosen. The most frequently mentioned tools (often used in training) are of three types: (1) Set of tools around collective action, such as the structure, rules, and procedures, and recruitment; (2) material stimulation tools, such as compensation, bonuses, promotions and, of course, the performance measurement related to them; (3) ideational or ideological influence of tools, such as vision or values. The tools available to managers can have effects retarded. In the case of compensation and performance measurement systems, the effects are almost immediate. In other cases, such as recruitment or training, the effect is felt later. In what follows, we develop just three of the mentioned tools or recruitment, remuneration, and promotion systems, and training. Indeed, the health of the organization is directly influenced by the recruitment of people it needs. Depending on the nature of the organization, the recruitment of effect can manifest itself more or less long-term or recruitment, remuneration and promotion systems, and training.

In a "red tape," recruitment has a relatively rapid effect since behaviors are routine and standardized. In a "professional bureaucracy" such as a university or hospital, the behavior is directly related to the nature of recruited professionals. In this case, the effect of recruitment is felt in the long term because recruitment is seldom massive and that those already in place dominate organizational life. Thus, the Japanese or Korean car companies that have invested in North America showed that it was possible to generate fast enough new behaviors through recruitment and initiatives such as training. By cons, universities, recruitment does not change behavior very slowly.

In the context of strategic management, executive recruitment is very important. Specialists in this field are a booming business to find the best people to fill senior management positions, people who can meet the challenges associated with the operational approach of the company concerned. That's why business leaders recruitment mandates always begin by clarifying the strategic context for this: it then directs research to a leader capable of realizing the operational approach of the company to modify or change it. Executive recruitment, when done well, so aligns the objectives from the start of the most critical people in the organization with the objectives of the organization. Members of a board of directors have a responsibility to ensure we choose leaders who share the great values and direction of the organization, and will therefore be able to translate them into consistent operational choices with these values.

Also, pay and promotion systems have an immediate impact on people's behavior, and thus the organization. These effects are not deep, although in interaction with other systems, they can influence the organization sustainably. The remuneration system is the basic system on which is built the exchange between the organization and the individual. The individual is supposed to cooperate, in exchange for which it receives material compensation. The compensation effect can, however, be disturbed by the influence of other stimulating factors. Thus, when the remuneration system is transparent, stimulation is related to the fairness of the compensation system, first within the organization and with similar organizations.

Pay systems specialists pay much attention to three key issues: the link between the work and remuneration, the meaning of the remuneration system, particularly with regard to justice and equity, and finally, other compensation components, such as performance bonuses, rewards associated with specific performance, and financial benefits, such as a car, subscriptions to clubs or associations, etc. Premiums are often divided into two parts: the first is granted if the executive achieved the objectives of its division, its function, or its affiliates; the second is granted if the company as a whole achieved its objectives. Here, we see the direct link that attempts to establish between the operational approach of the company and the compensation system. It is suggested to establish ways to monitor and measure progress against the operational objectives. It demonstrates that the compensation and performance measurement instruments are the backbone that ensures that the behavior of managers and employees is aligned with the operational approach.

In short, we can say that the design of the operational approach determines what should be done, while the reward systems ensure that the organization's members will work to achieve this operational approach. It demonstrates that the compensation and performance measurement instruments are the backbone that ensures that the behavior of managers and employees is aligned with the operational approach. In short, we can say that the design of the operational approach determines what should be done, while the reward systems ensure that the organization's members will work to achieve this operational approach.

People are sensitive to factors that have nothing to do with money or material goods. They need hope, ideal, explanation of the meaning of life, and relationships with others. The physical stimulation is illusory and that alone it can never completely motivate individuals. When basic needs (including physiological and safety) are reasonably satisfied, the best organizations are those that are able to persuade their members to contribute to their work has value in itself. Training is a tool that allows members of the organization to appropriate business goals of the company and the values that underlie them. Also, in a knowledge economy, staff training plays a very important role. It has two objectives: a strategic objective and a technical objective. In the first case, it is to socialize people, especially managers, to the purpose of the organization, its values, and its modes of operation. The aim is to promote the emergence and consolidation of a culture shared by all.

When individuals share the major objectives of the organization and the operational approach is consistent with these objectives, it is then easier for them to work on the implementation of this operational approach. In the second case, the objective is to develop and strengthen the skills needed to perform effectively and efficiently, the operational approach of the company. Larger companies often have training institutes in order to have control over what is passed on (taught) to members of the organization. They are also concerned, more generally, for managing organizational knowledge. The effects of training on the implementation of the company's operational approach are felt in the long term, but it is a crucial investment because it can give the company a competitive advantage difficult to copy.

The research to which we briefly referred shows the importance of a fit between the selected operational approach and the various structural arrangements of the organization. But the structure is not an implementation tool for the operational approach once it has been formulated. It also provides a framework for strategic action. Here, we focus on two aspects: first, the structure that influence and constrain the operational approach; second, the structure which provides the framework for strategic action in everyday life. If we consider the operational link structure approach, the operational approach forced the structure. But it is also true that the structure in established organizations can force the operational choices. This is what has led some authors to reverse the relationship proposed by Chandler and affirm that the operational approach follows the structure, rather than precede it. Today, we consider that the operational approach is influenced by the existing structure, but that a new operational approach, to realize, often requires new structural arrangements.

Desjardins illustrates the influence that the structure can have on operational choices. The cooperative structure of the institution of savings and credit, which gives local credit unions an important voice in the proceedings of the group, always constrained operational choices made by the leaders of this institution in its history: focus on individual loans rather than commercial, refusal to provide a credit card, noninvolvement in the cable, etc. The operational approach of Desjardins Group and the strategic actions of its leaders do understand that if one considers the type of structure, the structure affects the strategic action otherwise. The different forms of structure that we discussed are not limited to a flow-chart. A structure, it is not an arrangement positions. A structure is also an arrangement of management processes that have significant influence on individuals who occupy different positions in a given structure. It is through the combination, sometimes rustic, sometimes clever, management mechanisms that gives life to the structure of the organization. But there is more. In a logic of strategic action in everyday life, where we want the participation of all members of the organization for the training of operational choices, structural arrangements are very important. The strategic action every day requires special structural arrangements: ad hoc structures, temporary structures, transverse structures, project management, etc. It also requires management processes that enable stakeholder involvement in the formation of the operational approach. The leaders have a role very special to play in the establishment of these structures and the management mechanisms.

CHAPTER 14

Metadata and the Complexity Mechanisms

The realization of an operational approach requires appropriate management structure and processes. But to achieve an operational approach and reach performance, leaders must rely on factors other than those to which we have referred so far. In this chapter, we will pay particular attention to the organizational culture and leadership. Organizational culture and leadership, although controversial concepts are essential to achieving the operational approach. We will discuss their role in both the implementation of a planned operational approach and as a framework for action in the operational process.

In the first part, we focus on the concept of culture. In the second part, we will discuss strategic leadership by addressing the different ways that leaders can exercise. We first define what we mean by "culture" and we will mention the different sources of organizational culture. Subsequently, we discuss the role that culture can play in achieving an operational approach. The growing popularity of the concept of organizational culture coincided with the runaway success of Japanese companies in the international market.

Until then, the importance of the human factor had been considered mainly in terms of individuals and groups, from the perspective of the school of human relations. From that moment, the influence of the national environment on how to do the business on the sociocultural rather than economic has gained importance in strategic thinking and has attracted interest for the concept of organizational culture. But what does the organizational culture and can we manage it? Despite the controversy surrounding the use of the concept of culture, to which we shall return later, definitions of culture are quite similar.

The set of basic assumptions—that a given group has invented, discovered, or developed by learning to cope with its external and internal integration problems adapting—which worked well enough to be considered valid and taught to new members of the group as being the right way to perceive, think, and feel in relation to these issues. Culture comprises three interrelated levels from deeper and intangible at the surface and manifests. The set of basic assumptions is the deepest level of culture. It includes the assumptions, beliefs deeply rooted concerning the nature of reality, man, environment, etc., that unconsciously guide our perceptions and ways of thinking and doing. For example, belief that humans enjoy working or are fundamentally lazy.

These basic assumptions affect the second level, that of values. These, although they are often taken for granted, can be made explicit, especially when strategic thinking exercises. The valor of legislative character and drawn attention to what is considered important, on what is valued or not. They usually form a coherent whole, a gestalt, and a guide for action. According to Schein is by uncovering the value system of an organization that one can deduce the basic assumptions which themselves are inaccessible. The third level, artifacts, is the tangible manifestation of the values and basic assumptions. This level includes the social universe and equipment built by the members of the organization. The structures, systems, and organizational practices, as well as the products are all tangible signs of the culture of the company. Apple's case is eloquent in this regard.

Think of the Macintosh, the iPod and more recently the iPhone, whose output was accompanied by a whole page-media, it is clear that the products of the firm clearly express its culture, including values related to innovation, originality and aesthetics, and allow it to distinguish itself from its competitors. Similarly, the culture of McDonald's, based on standardization and speed, is recorded in the physical appearance of its restaurants, their predictable operation, and uniformity of its products. In both cases, a strong culture we see in synergy with the operational approach of the company as well as the products are all tangible signs of the culture of the company.

However, organizational culture does not emerge in a vacuum: the organization is embedded in society. On one hand, it is influenced by the dominant culture of the surrounding environment, national culture. Thus, Rieger and Wong-Rieger (1988) in a comparative study of international airlines, show the influence of the culture of the country of origin on the structure and operational approach of these companies. Based, among others, on the work of Hofstede (1980), they establish five cultural configurations according to the orientation with respect to power, power distance, analysis, and risk. Where culture is at odds with the core values of the company, the organization can have legitimacy problems that may affect the achievement of its operational approach. This may be particularly the case when companies internationalize their activities.

In addition, Walmart claims to have a corporate culture making unnecessary the use of union. The US giant has been the focus of considerable controversy surrounding the closure of a store unionization process, where the right of workers to organize is an important social value. Conversely, Ubisoft, the French video game company, embodies values such as creativity and playful spirit, which is associated with their culture. Social recognition is an important part of the identification of members to the organization and its operational approach, especially when companies internationalize their activities.

On the other hand, national culture does not influence alone the organizational culture. First, society is not homogeneous; we need only think of ethnic, religious, linguistic, and generational introducing diversity in the organization. The management of cultural diversity, of their AIL, a theme which is given more and more attention nowadays in organizations. Also, there is a culture specific to the industry in which the organization works. Some then speak industrial recipe (Spender, 1989) or cognitive industrial communities (Porac et al., 1994) to describe the ways of doing and thinking that characterize a given industry. For example, Porac and colleagues describe the business model that characterizes the high-end cashmere industry in Scotland and demonstrate how it is taken for granted by members of the industry and is a sustainable recipe that forced the behavior of shareholders. Similarly, the existence of different sectoral approach in banking and brokerage that is causing significant problems has experienced the Bank of Montreal during the process of acquiring brokerage firms Nesbitt Thomson and Burns Fry (Roch, 2003). Even after several years, it is found that there is little interaction between these different subcultures within the company.

Finally, occupational and professional cultures are another important source of diversity in organizations. These different organizational subcultures can be in harmony with the dominant organizational culture or, on the contrary, constitute against subcultures that undermine it. This is why some privileged strategic directions by the leaders, although quite rational from their point of view, to be rejected formally or informally by some groups of the organization on behalf of an alternative rationality. This is a common situation among other organizations in the field of health and that of the cultural field. Thus, when we want to implement a new operational approach in the field of health, professional logic often conflicts with the administrative logic. This was the case, for example, when introducing customers turn in the health network. In the arts, there is talk of two-headed organizations where the art direction and the administrative leadership must cooperate during the development of the operational approach. As can be expected, it is not without provoking internal tensions.

Generally, you have to report the existence of three subcultures that often conflict in large organizations: those leaders, operations people, and engineers. Leaders tend to focus on financial issues and competitors, operators of local concerns and focus on the people, while engineers are mostly interested in technology. It is not surprising that they have difficulty understanding and agreement. However, several authors emphasize the importance of developing a strong corporate culture that is to say distinctive, stimulating and widely shared, to support operational approach. Indeed, if the members of the organization share the same culture (if they are comfortable with the values and beliefs that underlie it), they will enthusiastically participate in carrying out the operational approach (to the extent that it is consistent with the existing culture).

But a strong culture is a double-edged sword. It can be a source of sustainable competitive advantage, because a distinctive culture is difficult to replicate by competitors, but it is also difficult to change and can become a source of inertia that prevents the company to adapt when the environment changes. If the journal provides a good example of a company that has developed over the years a so-called high culture that of a committed newspaper, intellectual rigor. The company culture is a reflection of the nationalist culture and it is, in part, that it is so widely shared by its members. This distinctive culture that has developed in line with the company mission is both the most solid base which underpins its competitive advantage and the policy framework that influences its strategic evolution. Indeed, as we have seen, this culture is forcing policy changes that can be undertaken by this newspaper in response to the changing environment. This brings us then to ask the question: How to manage the link between operational approach and culture?

The answer to this question has generated much debate. We can group the positions into two camps: there are those who see culture as a lever to achieve the operational approach and those who see it primarily as part of the strategic action. The former consider that to be effective, the organization must have a coherent organizational culture whose beliefs, values, and norms are widely shared by all members of the organization. In their view, culture can be used as a management tool and can help in the implementation of the operational approach. In this case, it comes essentially acting on cultural artifacts, such as symbols, language, structure and compensation, and promotion systems to change the behavior of members of the organization and make it more compatible with operational approach.

Thus, during the dismantling of AT&T monopoly imposed by the government, its leaders have begun a significant cultural change. To facilitate the transition from a regulated environment to a competitive environment, they moved the headquarters of the company, changed the name and logo, and implemented a series of measures (speech, video conferences, training, promotions, new partnerships, new products, etc.) to advise the members of the organization the new behaviors and appropriate values such as leadership, collegiality, and speed in decision-making. Furthermore, it is stated that the management has also stressed that these values were added to some existing values, including the importance of customer service and human resources, as well as fairness and emphasizing continuity in change.

The only cultural change that can be managed is incremental change, that is to say the introduction of new values that are not antagonistic to those of the existing culture. In what he calls a virtuous circle, members of the organization are encouraged to experiment with new behaviors that, if successful, then wind valued and idealized. They come in, over time, to be taken for granted and integrate with organizational myths. Moreover, the Cultural Revolution would actually destruction of the ancient culture. However, although it is relatively easy to destroy the existing culture by changing artifacts like structure and systems.

Other authors rather consider that the organization is not a culture that can manage, but the community of shareholders is in itself a culture. In this perspective, culture emerges from the collective history of the members of the organizational community.

The culture is then seen as a knowledge structure; a system of shared meanings and reflect largely unconscious processes. It is almost impossible to change intentionally and becomes instead the framework around which defines the operational approach of the business and even, from which it emerges. This is called dominant logic or organizational paradigm. Thus, explained the difficulties experienced by a firm of consulting engineers, with its diversification process by the existence of a dominant logic incompatible with its new activities. The company would not have managed to take full advantage of its acquisitions, because they were managed by engineers who did not understand the rules in the new sectors.

Another interesting example is the cooperative banks; the operational approach has evolved by adapting to organizational culture. The development of these structures was marked by the values of the founders. Some of their beliefs have become myths that there was no question; for example, the idea was to promote savings by discouraging the use of consumer credit.

For many years, no one dared challenge this way of conceiving things. Thereafter, the members face a different situation from that which existed began a strategic shift that resulted in the creation of the VISA Desjardins card. However, to achieve this change, which could be seen as inconsistent with the values of the company, they showed that credit cards could be defined as a payment card, since the majority of people used this way, paying off their balance each month. This conception of credit cards has helped make it consistent with the corporate culture. To show their commitment to the philosophy of Descartes gardens, the shareholders of the movement still continue to warn their members when using credit card-related problems seem to them.

But culture conceived as a framework is not a constraint to the operational approach; it is also the context which can emerge as a new approach. An eloquent example in this regard is that of Intel. Thanks to its culture of innovation and autonomy of the researchers, this company from a memory card manufacturer, has become so unplanned business in microprocessors and is now a leader in its field. At Intel, a qualified independent process consisting of local initiatives for middle managers and researchers experimenting with new futures exists in parallel with the formal strategic process called induced process. In this context, culture becomes a source of strategic innovation rather than achieving operational approach tools.

We consider culture as an implementation tool of the operational process or as part of the strategic action; the strategist should be familiar with the culture of his company. It is difficult for the members of the organization have the necessary distance to analyze their culture, since the assumptions and values are largely taken for granted, and therefore inaccessible. They then offer to call a drab former analyst who will help members of the organization to uncover their culture. The strategist must be familiar with the culture of his company. It is difficult for the members of the organization to have the necessary distance to analyze their culture, since the assumptions and values are largely taken for granted, and therefore inaccessible. They then offer to call a drab former analyst who will help members of the organization to uncover their culture.

The authors have proposed different approaches and developed various analytical tools of culture. Thus, the ethnographic approach, the clinical approach and auditing, which are based on various models, are part of the main approaches that have been proposed to analyze culture. The ethnographic approach is the most traditional approach to analyzing culture. It requires an observer to come live in the organization for some time to become familiar with its practices and to know its norms, values, and beliefs. As the anthropologist did with the tribes, it is for that person to recognize different cultural artifacts of the organization and to interpret them to remove the values and beliefs of its members.

Among the most frequently mentioned items include: rites (the events that mark the organization of life); routines (the ways of distinguishing); myths (stories and anecdotes about the highlights and important personalities that are frequently told and that highlight what characterizes the organization); the symbols (logos, slogans, and objects that have special meaning); organizational structures and systems, including recruitment, promotion, and control.

All these indicate what the importance is given and attention in the organization, what we value, but also what we reject. It is therefore not enough to list them, but rather to analyze them to understand the values and beliefs that underpin them. Thus, the structures and systems are reviewed here in terms of their functional efficiency, but according to their significance for members and their consistency with the organization's values. For example, the centralization of certain activities might be justified on the basis of the speed of decision-making and cost control, but be incompatible with the values of autonomy and sensitivity to local realities promoted by management. The structure, in this perspective, is analyzed,

First, it is to observe events and distinctive practices, to see what the structure puts forward and what it hides, which is controlled and which is, of note common expressions and design spaces, in short, everything that enables members of the organization feel at home and foreigners to feel the difference. Second, we must understand how and why the organization members interpret these artifacts, those to whom they give the most importance, those they identify with and which are rejected or treated with cynicism. Then we can take the pulse of the organization that is to say to what extent the culture is shared, and to uncover the existence of subcultures. Thus, after a long observation stay in the company resulting from the merger of Irving Samuel and Jean-Claude Poitras Design, Roll (1995) identifies significant cultural differences between the two entities, based on cultural and linguistic differences, as well as various designs of fashion (conservative vs flamboyant) and business conduct (emphasis on production vs on marketing). The existence of such differences is much hindered the integration of the two companies and partly explains the failure of the merger.

Based on cultural and linguistic differences (one Anglophone Jewish and Italian descent, and the other francophone), as well as various designs of fashion (conservative vs flamboyant) and driving business (focus on the production vs on marketing). The existence of such differences is much hindered the integration of the two companies and partly explains the failure of the merger.

This ethnographic work takes time and is not within the reach of every business. This is why other authors advocate the use of the clinical approach, based mainly on interviews with various company stakeholders. With the different perceptions (or homogeneous) of people working at several levels and in several areas, the specialist will look to see how the members of the organization analyze and understand the values and beliefs of the organization. It is through the analysis of convergent and divergent visions we discover the norms and values of the organization and those of subgroups. It is then possible to detect a dominant culture, if it exists, and the most important subcultures. By defining culture as the set of basic assumptions, it suggests a series of questions to help establish a cultural diagnosis. These issues are briefly mentioned in the following: relationship to the environment: you perceive yourself as a pioneer or follower, adventurous or cautious, etc.? Relationship to others: is it better to be proactive and tenacious or conciliatory and cautious, conservative, or optimistic, cooperative or competitive, individualistic or collaborative, etc.?

Relationship to the truth: the truth-it comes from the analysis, wisdom, experimentation, social consensus, etc.? Time relationships are you facing the past, present, or future? How is the time cut in your organization? Relationship to human nature: it is a human being generally well-meaning and reliable, or opportunistic and manipulative? Can it change, improve?

It also defines two major dimensions that allow to deepen this analysis: external and internal integration adaptation. Thus, for the external adaptation, focus is on the values that underpin the mission and operational approach of the company, the performance criteria, and the way to correct the problems. For internal integration, we will look at the language used (to uncover the frameworks, shared meanings systems), the nature of the boundaries between the groups in how resources are allocated (criteria, distribution, process, etc.), standards and how are gender sensitive issues like religion and ideological differences. This type of analysis allows to study, inductively, own cultural dynamics of each organization. Based on the idea that culture is actually built on the opposition or dilemmas, suggest that it is important in this analysis to decode the implicit oppositions in cultural choice. Thus, when evaluating the link between operational approach and culture, one is able to predict whether certain values that we would like to put forward to implement the operational approach conflicts with existing values and assumptions, and thus arouse significant resistance. Finally, some authors advocate the use of the audit, based on questionnaires based on pre-established dimensions.

This deductive approach, which is faster and cheaper, to measure the perceptions of members of the organization on certain issues identified by specialists. Thus, in the case of AT&T, this approach was used to explore perceptions of the members of the organization toward change, including new values that management wanted to put forward. However, this type of survey, though it has its usefulness, does not help to understand why people have developed these perceptions or the meaning they give to what they experience in the organization. Moreover, it does not expose subcultures that exist within the organization and can play an important role in the operational approach. This approach was used to explore perceptions of the members of the organization toward change, including new values that management wanted to put forward.

In conclusion, it is clear that culture is an important dimension of applied management. The analysis of cultural dynamics is to better understand the interaction between organizational culture and subcultures and their compatibility with the operational approach. The balance of management between dominant culture and subcultures dynamic is important for organizational performance. A dominant culture is necessary to give coherence to the operational activities, but the presence of dynamic subcultures promotes strategic flexibility. In the case of the implementation of a new operational approach, leaders can wind rely on the most compatible subcultures with their new intentional operational approach. For example, to commercial turn in some state companies, CEOs were based, among others, the groups of farms, who valued the service to subscribers and had a long history of domination equipment group, which itself, focused on the construction of dams.

Moreover, if we allow subcultures to express themselves, they introduce in the organization of different perspectives that can contribute to emerging way to strategic renewal. Thus, the Notre-Dame Hospital has become a leader in palliative care, both in terms of research and in terms of training, through professionals (doctors and nurses) who managed to change attitudes and develop palliative care in a hospital environment dedicated primarily to curative care. Leaders must always be aware that the development of a coherent culture and its coexistence with different organizational subcultures pose significant management challenges, which require the presence of an enlightened leadership. This leadership can be exercised in different ways, as we shall see in the next section.

The organizational culture phenomenon that we looked up to now is both a tool to be taken into account in the implementation of the operational approach and an environment affecting the strategic action in everyday life. This culture is always the fruit of a long maturation process. Easily we cannot forge it and can never be changed easily and quickly. It is a different leadership. The literature in management and operational approach contains many examples of companies whose rapid development is linked to the arrival of leaders who exercised a strong and convincing leadership. If the role of leadership in operational strategy is not in doubt, it is quite different from how to design it. First, we define what we mean by "strategic leadership."

Second, we show how this leadership can be exercised in companies and what its importance in achieving the operational approach is. This view management has often contributed to a devaluation of the role of manager and an apology for the action of leaders. No leader wants to be associated with a person concerned about the trivialities of daily life, incapable of intuition and manager with a niggling and technocratic rationality. And this is understandable! But the life of organizations and study of great managers lead us to reject this dichotomous approach to management. A leader must be both a manager and a leader; he must manage the present and be able to design the future; he must be concerned about the short term and an interest in the long term; it must ensure the stability of the company and be able to change it.

Sometimes it is not a single leader who combines the managerial skills and those of leader, but different members of a management team. Then we find within this team complementarity which allows the well to grow business. The company studied had significant problems when the technocrats have replaced the artists and artisans in the company's head. The management team was no longer able to develop a long-term vision of the company. The operational approach of the latter was no longer defined by the investment rate of return in the short term. The financial reasoning was substituted for strategic thinking. In an organization, leadership can take different forms. Here, we focus on operational leadership, which is to say to the leadership of the leaders in the operational approach, both in its formulation and in its implementation, but also in its implementation in everyday life. Some prefer to speak of visionary leadership.

Leaders are able to have a vision of the organization, but are also able to communicate that vision and to share with other members of the organization. They communicate and often use metaphors, analogies and symbols to explain their vision and make it understandable to the rest of the organization. This same process is described as leaders who have two important roles: they have to "make sense" and they must be sense giving. The operational leadership is often associated with major changes, but it would be wrong to believe that it is only important when an organization has to make such changes. Strategic leadership is an important asset for the operational approach, whatever the adopted operational approach, and leaders must be able to exercise it.

The leader must be both the architect of the reason, the designer of the operational approach, and creative context. These are three great ways to exercise operational leadership. As an architect of the rationale, the leader cares about the values and principles that guide the organization. The role of the leader is crucial. It is his duty to transmit values within the organization, and it is he who must ensure that these values preside over major decisions of the organization. When an organization is imbued with strong values, these values are reasserted and defended in time, and acquires a strong social recognition, we speak of institution.

By comparing businesses in the same industry, requires that companies with a pragmatic idealism, that is, to say a few core values and a concern for economic efficiency, are more efficient and better developed than those concerned than by profit. The values they have identified may be of different types: contributions to society with useful products (Merck), concern for the needs of all stakeholders (Johnson & Johnson), or development of the best of each employee (Sony). But what matters to these authors is that leaders ensure that these values are enshrined in the mission of the company and the shares of the latter are consistent with the values enshrined in the mission. This concerns both the type of implanted structure.

But the leader can exercise leadership in a different way, being the designer of the operational approach. This is the most classic to consider exercising operational leadership so. At first, the officer working on the formulation of the operational approach in trying to establish a fit between opportunities and threats of the business context and the strengths and weaknesses of the organization, taking into account its own values and of its social responsibility. To accomplish this daunting task, especially in complex organizations and large, the officer may be assisted by planners or industrial economists. But it is the leader, and he alone, has the task of synthesizing and ultimately to the strategic choices required.

When the operational approach is formulated, the leader then uses its communication and persuasion skills so that it is implemented with the least possible distortion by the rest of the organization. It therefore ensures the implementation of the operational approach formulated. He is the guardian of the values of the organization and ensures that it stays on course. It works, when necessary, as a mediator and negotiator between different individuals and groups. It puts in place structures and reward and punishment systems to ensure satisfactory implementation of the operational approach. This is an important way to exercise strategic leadership with the least possible distortion by the rest of the organization. It therefore ensures the implementation of the operational approach formulated.

It is the operational approach that has revitalized GE. GE was a diversified company in several business areas, and leaders understand more clearly what this business, despite the structure of strategic business centers and strategic planning process that had been set up. We had to address this complexity. It was defined corporate operational approach from three circles (the original business, high-tech activities, and service activities) and it stated two clear goals: to be number 1 or number 2 and have an annual growth rate of 15%. From there, he changed the company structure and management process, including the planning.

Apple also recognizes the merit of having been able, at different times in the life of the company, to develop a vision that not only made the business success but also redefined the rules of the industry. This role of the operational approach of designers is a very demanding role. It is necessary that the leader has the vision, that is to say the ability to imagine the future and project the organization in the future. It is also the perception of the business context and organization is the least distorted possible through abnormal cognitive biases; by inadequate categorizations; by an oversimplification of the context of business and organization; by obsessions, conditions, or dysfunctional personality profile. Finally, the leader can exercise strategic leadership of a third way, being the creator of the context.

It is then no longer defined as being solely responsible for the formulation and implementation of the operational approach. It is defined as one whose job it is to set up an environment that will enable the participation of other members of the organization to the operational approach. The case of the Oticon business is very interesting as it allows us to understand both the roles of leaders and the role of other shareholders in the organization of the operational process. The Danish company Oticon is a world leader in hearing aids. The study of strategic initiatives that have characterized this business over time brought the two authors to design the operational approach as a "guided evolution."

The members of the top management group had five responsibilities' at hand: (1) to develop and articulate strategic goals qui defined the strategic intent of the organization; (2) to sponsor strategic initiatives; (3) to allocate financial capital to strategic initiatives; (4) to recruit people to the organization; and (5) to take responsibility for the development of one area of functional expertise and knowledge in the organization. As seen, executives have an important strategic leadership as their responsibility to make clear what the operational objective of the company is. They are the architects of the reason, but they are also the creators of context because it is their responsibility to set up the context for the development of various strategic initiatives.

In practice, this is to encourage all people to start a project and assign to these projects the necessary human and financial resources. The significant role played by leaders does not drain that of other players in the organization. Instead within the leadership set by a firm's strategic intent, celebrities have some leeway in deciding qui external opportunities to pursue and qui forms and combinations of human and social capital to preserve and develop. Over time, this freedom may result in the actual decision-making and resource allocation processes exercising some "stress" on the strategic intent. This view of operational leadership has several implications. On the one hand, the leader is no longer considered the only actor involved in the formulation of the operational approach.

There are other people in the organization who play an important operational role, and it is up to the leader to ensure that the context allows these individuals and groups to play their role. It is therefore a much more collective and participatory approach to operational approach. Moreover, the operational approach may well not be made before the action, but be defined during the action. The leader must ensure that the context makes possible an operational action daily to the various shareholders of the organization. It is therefore a much more collective and participatory approach to operational approach. Moreover, the operational approach may well not be made before the action, but be defined during the action.

Moreover, as the operational approach is not than intentional and that may emerge during the action, the leader must ensure that the organization structures and management modes that enable it to identify these emerging operational approach. Given that the literature on this third type of leadership is much more recent and it is based on new approaches both in organizational theory and operational approach (mainly constructivism and structuralism), given as the difficulties of the leaders to make only relevant operational approach in situations of complexity, so it is this third type of operational leadership that we will devote the following pages. We will show that leaders who are context creators are relying on some strong values, by setting up appropriate structures and developing appropriate management. When people cannot perform some complex tasks, they are forced to work with others.

What are the limits of people, both cognitive and physical, leading to define the organization as a system of cooperation? It must be said that without cooperation, there was no organization. And add that the role of managers is to get individuals to cooperate for the achievement of organizational objectives. We must highlight the fact that the cooperation of members of the organization is essential for effective and efficient production of goods and services. This is a design very "functional" of cooperation, which is often valued only at the implementation stage. However, cooperation in the organization can go beyond that of the execution and implementation of plans and programs already formulated by management, and is seen as very important for the operational choices.

Thus, the organization's members cooperate when they perceive that their contribution is offset by a fee they consider fair. This compensation may be hard, but we know that the material compensation cannot, by itself, ensure sustainable and economically viable cooperation. It is therefore necessary that cooperation is also induced in more intangible mechanisms "persuasion," and persuasion is often possible when trust between leaders and groups in the organization. To establish cooperative relationships in a context of uncertainty, confidence would be more effective than coercion or rational calculation that involves a rigid and expensive control system, and therefore inefficient. The trust would be perceived by the agents as a modality of action, an arrangement leading to cooperation. Trust acts as coordination mechanism between agents to remove uncertainty about the expected results of a relationship.

Leaders must work to build that trust, every day, over time, in different decisions, operational and strategic, they are taking or are taken by the other managers of the organization. Trust is often fueled by symbolic actions of the leaders. When available, the trust allows members of the organization to work effectively to implement operational practices and make significant changes. The existence of trust relationships within organizations can reduce the intensity of political games, but it cannot evacuate.

The stakeholder community is not a homogenous whole, oriented toward a common goal. It consists of individuals and groups who may have divergent or conflicting interests. One has only to think of the interests of the leaders, which may be different from those of shareholders and the interests of these two groups in relation to those workers. Think also to the divergent interests of different professional groups (doctors, nurses, and paramedics) who work in a hospital; crystallize their interests when it comes time to discuss strategic orientations to the hospital system but it cannot evacuate.

Within the community of shareholders, power and influence are unevenly distributed and are based on assets whose shareholders have. These assets can be of all kinds: sex, ethnicity, education level, professional

competence, capital, control of a significant uncertainty for the organization, recognition outside the organization, place in the hierarchy, privileged
links with the context, etc. Being a white man, having a university degree
(in engineering rather than anthropology), that certain power bases that
give those who hold a place and a privileged role within organizations, and
therefore in the process of decision, optimal decision. It was emphasized
the dynamic nature of power relations in the 100 largest US companies.
The dominance of entrepreneurs and production staff, we went to that of
sales and marketing staff, and finally to the people of finance. With these
competing interests and power unequally distributed within organizations,
so it is not surprising that we are witnessing political games of all kinds.

This is an unavoidable reality of organizations. Very often, in order
to increase their ability to affect the action and impose their vision on
the organization, individuals form coalitions. These coalitions can be very
effective. How dominant coalition could have enough power to change
the mission and goals of an organization and to uphold its own definition
of the future of the organization? When we created some hospitals, they
had a priority to prevention rather devote their resources to the curative
aspect, as was the case in most hospitals at the time. The project should
be based on general practitioners, not medical specialists. He must finally
give emphasis to health care professionals other than doctors and nurses
(psychologists, physiotherapists) as was the case in most of the hospitals
at the time.

This mission contrasted with the traditional mission of a hospital and
was to redistribute power between the various stakeholders of the hospital.
In doing so, she questioned the place, role, and privileges of the specialist
physician. For a subtle coalition with nurses, specialists have managed
in less than 5 years to change the mission of the hospital and to ensure
that their interests would be preserved in the new hospital. Leaders cannot
eliminate the formation of coalitions and political games. But it is up to
managers to monitor these political games so that the organization does
not turn into a political arena.

They must ensure that we do not attend a politicization of the functioning of the organization. They must use their leadership so that the
objectives of the organization are needed to confront specific objectives
of certain individuals, groups, or coalitions. The participation of different
stakeholders of the organization to operational choices cannot be done
if there are structures that enable such participation. The operational

approach in large companies often emerges projects defined by certain individuals and groups working on the operational side of the business.

These innovative projects are subsequently supported and defended by intermediate management levels that are promoting with the management team. Also, the existence of operational subsystems, dealing with particular issues (acquisitions, expansion, government relations, etc.) allows different subgroups of the organization to participate in the determination of operational measures. The role of leadership at the top is then to define these operational subsystems and to coordinate their action. This is from the action of different subgroups, and their coordination, which forms the corporate operational approach of the company.

It takes a lot to emphasize the establishment of structures that allow diverse groups of the organization to play a role operational choice. These can be committees or transverse structures that bring members of the organization to know, to exchange, and find solutions to operational or strategic problems. In this sense, project management can be a structure and a management that enable the participation of different shareholders in the organization. The establishment of communities of practice can be another way to promote the participation of individuals in the formation of the process.

For such a system to work, it is necessary that the structure operationally allow the emergence of innovative projects. It is also entrusted to middle management an important role in the promotion and defense of these projects. It should also emphasize the need to set up structures and management modes that enable them to play that role. Finally, we must accept that the leaders are the architects of the purpose and the creators of the context, but they are defined more as the only developers of the operational approach itself. The management team has to continue to play the important role of "guardians of values" of the company, to define "the playing field" to promote a participatory approach and to establish a context for creative participation.

For instrumentalists, the reality is too complex and rationality leaders too small to operate in the classical models of strategic analysis and planning. It is best to deal with the context of business and organization incrementally, in small steps. This is a "disjointed incrementalism" open to organized voluntary, or better, it a "logical incrementalism" which is based on the existence of strategic subsystems we talked, and which attaches great importance to flexible coordination by leaders guided coordination objectives. These leaders have a general idea of the direction the company should take; they evaluate the initiatives of the various subsystems, accept or refuse incrementally according to events and opportunities. Thus is forged corporate operational approach of the company.

Operational incremental approach is based on various processes, including that of learning. Organizations learn, but they do so through individuals who learn and thereby creating tacit or explicit knowledge. These individual learning eventually constitutes the memory of the organization, that is to say, a shared mental model. When we talked about the culture, we refer to these shared mental models, which can be associated to directories, the "dominant logic" of the organization or the "organizational paradigm." This memory is manifested, among others, in the form of artifacts, structures, and routines.

The learnings that are members of an organization are of several types. We are sensitive to functional learning, but much less to strategic learning. Or, they can be critical to a company to be competitive. In a knowledge economy, such as that to which we belong, effective management of learning and knowledge can be the source of a strong and sustainable competitive advantage. The leaders then have an important role to play in the establishment of an environment that enables and facilitates strategic learning by members of the organization. A context for learning is characterized by discipline, overtaking, the trust and support.

But organizations also need to unlearn when their ways to allow them not to be competitive. Leaders must ensure that they have organizational practices that allow the questioning of prior learning and promote "unlearning" when necessary. Unlearning is usually easier when an organization is in crisis, but it is a situation that organizations seek to avoid. The challenge of leadership is therefore to ensure that their organization is able to unlearn, even in periods of relative stability, when the environment changes or the competition changes.

A longitudinal approach on the evolution of cooperation in strategic alliances shows that collaborative projects that have successfully passed through the learning cycles, reassessment and readjustment, while failed projects were characterized by little learning, by divergent learning, and an inability to adjust behavior, which led to much frustration. Honda's operational approach on American soil, illustrates how members of the organization, who were not at the top of the company, have been instrumental in the operational formulation.

Honda thought displacements 250 cc and 305 cc were the most likely to be successful in the US, but these models have quickly seen major failures, related to how to drive the Americans. Until corrective action is taken at the initiative of field representatives, Honda has decided to sell 50 cc engine capacity that the company did not consider it appropriate for the US market. The success was resounding. A new operational approach has therefore emerged from the action, and Honda then used his power to appropriate this market with distribution networks and a completely different mode of operation of those it planned out.

So the operational choices that emerge during the action, and then it is important that the organization has the means to find them. This is the new role given to planners. In traditional approaches to planning, planners' role is to analyze the context of business and organization, to synthesize and make strategic choices that flow from their analysis. We must remember that, planners revert to what they should always have been, namely analysts responsible for "operational planning." They must "work around operational action."

They then have to play various roles, including that of "acting as discoverers of the operational approach." An important role for planners willing to think beyond the planning may well is to discover emerging operational approach in their organization (or in the activities of competing organizations). After emerging operational approach are discovered, planners can get a better strategic control over them by assessing their viability in the same way they do for deliberate operational approach. Planners must therefore be on the lookout for the operational option that emerges during the action. This work is important because it allows the operational approach that emerged slowly to be identified and codified.

Acknowledging the emerging operations is therefore not to eliminate the position and role of deliberate choice. However, it is to say that the operational approach in the business can take different paths that planners must be sensitive if they want to truly help the company to be competitive and efficient. It is up to leaders to ensure that planners and analysts they surround themselves have the training and skills to truly play their role. As we have seen, culture and leadership are two closely related realities. These are the leaders who transmit values within the organization and ensure their protection. Culture and leadership are therefore very important elements for understanding the behavior of an organization and business processes that take place there. Competitive advantage is often due to the particular ability to converge and work together. He comes from a general willingness to cooperate on the part of individuals, each bringing a unique and creative contribution. Leaders can allow the organization to shape the organizational culture that competitors will struggle to replicate.

CHAPTER 15

Optimizing Enterprises Managerial Innovation

Concerning the reasons for the dominance of some companies in their respective industries, it is important to note that the development of these firms had nothing to do with the invisible hand of the market, but everything to do with the visible hand of the management. They are firms that have been able to solve the new problems posed by managing the incredible growth of networks of production and distribution that have become more complex and difficult to manage. Managerial innovation, rather than products or finances, is critical to business success in a given market (Chandler, 1986). Whereas, the activities of single unit of traditional companies were monitored and coordinated by market mechanisms, which produced and distributed the units.

Within a modern business enterprise are monitored and coordinated by middle managers. Top managers, in addition to evaluating and coordinating the work of middle managers, took the place of the market in allocating resources for future production and distribution. In order to carry out functions' thesis, the managers had to invent new practices and procedures which, in time, became standard operating methods in managing the generation and distribution.

As more sophisticated technology developed and markets expanded, administrative coordination replaced market coordination in an increasingly larger portion of the economy. By the middle of the 20th century, the salaried managers of a relatively small number of broad mass producing, wide mass retailing, and broad mass transporting enterprises coordinated the current flow of goods through the processes of generation and distribution and allocated the resources to be used for future production and distribution in major sectors of the American economy.

By then, the revolution in business had begun. For instance, the history of Standard Oil of New Jersey showed how the multiplication of activities led to difficulties increasingly large that required original adaptations to the functioning of the organization. In particular, to survive the multiple crises resulting from the growth, the company first had to focus on centralization facilitated by the functional structure, to prevent the dispersion and to increase efficiency.

However, centralization, that provided more efficiency in the beginning, became a problem later, when the activities were so numerous and diverse that the local initiative was needed. This then gave way to a major decentralization made possible through a decision structure by product or project. In the case of Standard Oil, the division approach was then accompanied by numerous innovations, including the creation of several committees' side to facilitate coordination. Such operation left room for the initiative of entrepreneurs.

The reorganization of Jersey had followed the operational action. But, the reaction was slower, more hesitant, and less decisive than General Motors (GM). This difference is partly due to the fact that Jersey had the most difficult problems. Jersey was both a centralized group, with functional departments, such as DuPont, and an association with no specific links, and ultra-decentralized, such as GM before the reorganization. Further, if it were to create operating divisions like DuPont, and general management like GM, then both had to be created at the Jersey, and revised, in addition to the organization of certain functional departments. In order to effect these reforms, Jersey had (in addition) to get rid of many of strong established (toxic to business) corporate culture. The "explosives company" was less instrumental in this success. On the other hand, Pierre du Pont was the great architect of this success. It should be noted that the principle of management by committee and a tendency to neglect organizational problems were part of this corporate culture that had to be gotten rid of.

Teagle and his staff waited long to upset these traditions and to adapt the structures to the operational action. We must attribute this shift to the personality, training, and activities of the leaders of Jersey. Except Sadler, Howard, and perhaps Clark, they thought little in terms of organization. They ran daily, and neglected the long-term problems, preferring action to analysis. That was why the reorganization of Jersey was made overnight without following a predetermined plan. It, by cons, adding a few other oil companies, if it were, addressed the organizational problems in a more rational way than Jersey. Standard Oil was a unique company because it was composed of a multitude of shady entrepreneurs, concerned about their independence, who had at heart the success of their part of the business and at the same time had the desire to maintain the integrity of the whole.

Decentralization could, in this case, be seen as natural, since the company was in fact only an association of its constituent entrepreneurs. Moreover, when asked to Rockefeller, how he explained his success, he replied with a terse formula that showed the importance of all these entrepreneurs they managed because they were able to share.

In fact, this profit sharing is especially accompanied by a division of responsibilities, which left a lot of initiative to sectoral or local officials. But, more importantly, it facilitated the company's management. It had become so large and diversified that the leaders at the summit could not understand enough to act wisely. The decision had to be local, with a minimum of overall coordination. This arrangement was simplified within the overall management of the company so that reasonable decisions, if not perfect, could be taken. This is the first reaction one should have with regard to the complexity: simplify to function. In this regard, the history of Standard Oil is not unique.

With the proliferation of activities, we must recognize that the difficulties of management have gradually increased today. Producers of food, medicine, and other brand of products in packaging, manufacturers of electrical and electronic equipment, and all kinds of engineering companies showed the ability to use their internal capabilities in order to outperform the competition by penetrating the markets of their related products. However, the same success of these growth strategies, facilitated by the multidivisional structure, created new challenges. The rapid expansion in branches in distant relationship with that of the company, or even unrelated to it, exerted enormous pressure on the multidivisional structure. It led to a breakdown in communication between the managers of the central staff and operational senior divisions.

This dysfunction had two causes. First, leaders of the headquarters often had little knowledge of the diverse technical processes and markets of the many companies they had bought or they had little experience. Then, the acquisition of more divisions simply created an overload for decision-making at headquarters. Before World War II, the staffs of large diversified firms and multinationals rarely managed more than 10 divisions, and only the giant firms were managing about 25. In 1969, there were 40 companies that administered many divisions and some even more.

Senior executives from headquarters, unlike their predecessors, did not have the time required to establish and maintain personal contacts with the heads of the divisions. They had no more detailed product experience, which was needed to evaluate proposals as operational managers to monitor their performance. Overloading was not the result of any lack of information, but the quality of information and the ability of the most senior leaders that needed evaluation. Indeed, they were beginning to lose the skills needed to maintain a unified company. Note that in this case, the whole is more than the sum of its parts. The complexity of the situation was just that the executive action capability was reduced significantly.

Without getting into technical definitions, we will retain that a situation is complex when: activities and technologies are so numerous that leaders can all understand, but, in general, they have a limited understanding of the resulting system; power is shared and dispersed so that these leaders have little constructive power to bring the organization in the desired direction.

The question that arises from this situation is this: how can we run a business when, simultaneously, it is not clear what is going on and we have not all the required power to act? Here, we will concern with some tools that are useful to simplify the task of the leader, including decision-making for resource allocation. The first part will be devoted to methods of analysis interested in the content of decisions. The second part will address the process by which decisions are made and how these processes can be modified to get the desired results. We are concerned here only with methods that simplify the problems of decisions faced by executives in large complex organizations. Essentially, these problems are related to the distribution of resources among the different activities of the organization. Traditionally, the decision of resource allocation was dominated by investment appraisal procedures recommended in the financial analysis textbooks. These can be summarized as follows:

- assess the flow of funds, especially funds for expenditure necessary to achieve the envisaged investments, and cash inflows that the investment is supposed to result in the future;
- calculate the present value of inflows and outflows, using as discount rate the rate of return required of the company by the market (Christmas, 1989); make the balance between the updated inputs (positive) and outputs (negative); the result is the net present value of the proposed investment;

- choose investments that produce the highest net present value or profitability of the investment (ratio of net present value and the total capital expenditure) the highest.

This type of assessment, which still dominates the investment procedures of firms, is very sensitive to the quality of projections of future results, but it does not make an informed judgment on these projections. Companies like GE were then forced to maintain a large number of skilled employees at headquarters to monitor the quality of evaluations by the operational divisions. More importantly, these methods do not allow to understand how the competitive position of the activities, which the company was subjected to, would be affected by the investment concerned.

This led the consulting firm McKinsey to develop a product portfolio model that met the needs of General Electric (GE). Most of the major operational management consulting firms have subsequently created their own version of this model. Items in this section supplement the general presentation of portfolio analysis, covering the operational choices. What determines the complexity lies precisely in the diversity of activities. But, the activities are not always easy to distinguish each other. The definition of the activities is in itself an operational decision. Thus, the railway companies had missed a rare opportunity to dominate the market for mass transportation of the 20th century, notably by road and air, defining themselves not as goods and passenger transport companies, but as the "rail transport companies."

This narrow definition caused the decline of these companies throughout this century. The reverse is also true. It can be defined so broadly that we may not be able to excel in all activities that implies. Thus, no company would be defined today as "Computer Company" because it does not want to say much, given the large number of possible activities in this sector.

Defining the strategic activities is then a key step in the operational approach, when in complex situation. In general, a strategic activity can be defined as an activity (or a set of product-market couples) fairly autonomous, for which we can define a strategy. This implies that we can clearly identify the competitors, and the resources used (equipment, staff, etc.) can be clearly separated from other activities. Many rules and guidelines have been proposed to facilitate this, what is generally called "strategic segmentation." Recall that the model BCG (Boston Consulting Group) focuses on strengths, in terms of "market position" and "attraction market," the product portfolio, markets or strategic activities of a company.

The key to the model is the relationship between the net cash flow (cash flow) of strategic activity and its characteristics in terms of market share or market growth. The goal is to acquire the portfolio that will produce the cash flow, which becomes higher and more stable over time, taking into account the patterns of each of the activities in this field. Using the model portfolio of activities, the manager then aims to maximize the strengths of the company, by balancing the production and use of its funds.

The central proposal indicates that in most competitive situations, there is a strong relationship between relative market share and market growth on the one hand, and the characteristics of the production and use of funds for operational activities, on the other hand. Thus, when a company makes, for a given activity, a strategy that allows an increase in the volume of business faster than its competitors cost advantages result. This cost-volume of business is justified by what is called the experience curve.

The BCG model describes the experience curve as a predictable relationship between the unit cost and the cumulative production volume. Building on previous work, made especially by the Army, US Air, and the learning curve in terms of production, BCG proposed that the learning effect extends to all the factors involved in value added, such as capital, labor, and fixed costs. The reasons for applying the experience curve are not known precisely, but several outcomes are mentioned, including: improving the efficiency of the workforce; the effects of improving methods and the introduction of new production processes; new designs and improvements in product design that allow saving raw materials; improved manufacturing efficiency or the use of less expensive resources; the effects of possible standardization of the product.

The curves of experiments are often presented on a graph where the x-axis usually represents the cumulative production volume, while the unit cost is represented on the ordinate. Because of the experience curve, the competitor who has the share of higher relative market activity (or product) is the one that has the most important cumulative volume and, therefore, the lowest unit cost. He thus generates the highest margin for a given market price. If the production fund is clearly related to the experience, and therefore the relative market share, use of funds, particularly for investment, of course depends on market growth. Maintaining or increasing market share in a growing market creates capital requirements that are even greater than the growth rate and thus the gains in market share referred are high. The dynamic between the production and the use of funds permits the classification of the activities to facilitate strategic decision-making, among others and also to facilitate the allocation of resources among different strategic activities.

When an activity has a strong relative market share, thus generating a lot of funds, and in a growing market, thus uses a lot of funds, we have an ideal situation. Indeed, not only we have the opportunities for growth, but also we have the resources to exploit them. Activity in this situation is called "star."

When an activity has a strong relative market share, thus generating a lot of money, and is in a low growth market, we do not use a lot of money, it becomes a net capital provider that become available to other uses. Activity in this situation is often called "cash cow."

When an activity has a low relative market share, that does not generate much money, and is in a strong growth market, thus requiring much capital, if only to keep that, it is in a delicate situation. It can only progress if we decide to invest money there, sometimes in large quantities. Activity in this situation is often called "question mark."

Finally, when a strategic activity has a small market share in a low growth market, we are facing a situation that people at BCG see as problematic and why they tend to suggest to abandon this activity. Activity in this situation was generally called "lame duck," which is unfortunate, because it would be more accurate to say that these situations, on a closer look, probably have to redefine their position or make an adjustment. The divestment should be considered only after a more detailed analysis.

It is desirable to take the funds produced by the cows to invest in the most promising question marks to make them stars, which, over time, become cash cows. A balanced portfolio should then include activities in the three boxes "virtuous," the box of "lame ducks" to be empty. When McKinsey made the diagnosis of GE, its consultants had recommended a similar portfolio management to that of BCG. However, GE executives were uncomfortable with the quantitative nature of the decisions implied in that model. They preferred among others to follow more sophisticated judgments about the competitive position of the business, rather than relying on the relative market share; they preferred also to think more deeply about the attractiveness of the market, rather than considering only the market growth rate. However, in order to perform a systematic approach, they decided that these decisions would be guided by a set of dimensions.

The competitive position and the attractiveness of the market for a given activity are then represented, as in the model of BCG in a matrix diagram. The recommendations for allocating resources resemble those of BCG, with a classification of activities along the selected dimensions. What then of the benchmarking model? The Marketing Science Institute has developed a database and a methodology that helps to clarify the competitive position of business as described in the model portfolio.

The methodology, called Profit Impact of Market Strategy (PIMS), compares the performance of a given activity to those of some 2700 units of the database, replacing the data on this activity in one of the derived regression equations empirically from bank data. PIMS thus produce four types of books, by substituting the data about this activity in one of the derived regression equations empirically from the data bank:

(1) a book of "Par," indicating how are the return on investment (ROI) and the net cash flow (cash flow, or FM) compared to RCI FM and similar activities;

(2) a book of "strategic sensitivity" by suggesting scenarios how ROI and FM can change if we change the characteristics of the model variables;

(3) a book of "optimum strategy," which aims to determine the combined decisions that maximize ROI and FM;

(4) a book "Lim," which allows a selective combination of the results of Par and strategic sensitivity.

The model is based on mathematical equations that are supposed to explain over 80% of the variability in profitability and net cash flow. The variables can be grouped into five main groups:

(1) the characteristics of the business context, including the growth rates in the short and long term, inflation of prices and costs, frequency of purchase, the number and size of users and buyers;

(2) the competitive position of the business, including the part of the market served, the relative share, the relative quality of products, price, the marketing effort and activity of new products;

(3) the structure of the production process, including capital intensity, the degree of vertical integration, capacity utilization, productivity, equipment, and persons;

(4) discretionary budget allocations, including budgets for research and development and marketing budgets;

(5) strategic actions, including changing patterns in the mentioned controllable variables.

PIMS uses two types of regression, one to 37 variables for the prediction of RCI and the other 18 variables for the prediction of FM. The main feature of PIMS is that it can replace many subjective assessments of the McKinsey-GE model derived by feedback of empirical data. Like most, important strategic variables, that may affect the performance, were introduced in the PIMS model; the differences between this and the observed performance can be attributed to non-strategic variables such as operational management.

Models such as BCG and PIMS, and those derived, have been designed to streamline decision-making when the complexity is too high and is testing the cognitive abilities of leaders, especially their ability to understand organizational phenomena and act effectively. The simplification proposed in the product portfolio model is actually a simplification of the strategic analysis. Instead of forcing the leaders to gain knowledge of a considerable amount of information on the strategic situation of each activity before taking a decision, the model digests this information: first, by reducing the two parts of the strategic formulation or internal analysis and external analysis with dimensions that can be examined and evaluated in a systematic way by collaborators; second, by offering "standardized" decisions to maintain portfolio balance.

Thus, the internal analysis is replaced by the relative market share in the model of BCG, or clearly explained in other dimensions' replacement models. Similarly, the external analysis is reduced to the market growth or dimensions that can facilitate the systematic assessment of the nature of the context. The model then suggests decisions that are based on logic. As leaders are in control of financial resources, we suggest they divide so that the portfolio of strategic activities is balanced. They must therefore ensure that the company's activities include producing activities funds today (the cows), that is, the dominant activities in their fast growing market, and generating funds for the future.

The leader only has to confirm the result of the analysis by affixing his seal to a decision after the technical work. It must also ensure that the portfolio is maintained in a state of dynamic equilibrium, that is to say, anticipate problems that could jeopardize the balance or, when the balance is not achieved, take decisions to achieve this. Thus, for the company Harlequin, a leading global public love novels publisher, his biggest strategic problem in the 1980s was to find activities that could gradually replace the company's successful activities. Over a period of about ten years, and despite several acquisitions, the results were rather unfortunate: it has hardly managed to do it. Simplifying the task of the leader resulting from portfolio model is significant, which explains the popularity that this instrument has experienced in media consulting strategy.

The model PIMS also proved outstanding as a complement to the portfolio model. It allows to document the competitive position of the business, and thus makes it a more convincing classification resulting in the model. Running a large complex organization provides a return amount to manage the portfolio of activities it carries out.

In terms of disadvantages, it must be said that most of the problems come from the same power of the model and the comfort it provides to decision makers. The analysis is so convincing that the leaders feel that the results will follow. This often results in decisions based on stereotyped and dangerous behavior. First, the terminology itself can be a problem for the company's management. Indeed, the classification generates behaviors that tend to confirm the correctness of the chosen term. Thus, the activities staff who have been classified as "stars" will have behaviors that require compliance related to this status. Similarly, the staff of classified activities "lame ducks" will be demobilized and thus confirm the "prophecy."

Moreover, leaders tend not to use their judgment and also do not encourage the judgment of their employees. The result of the analysis can become a kind of dogma that everyone will strive to meet. The model then becomes a kind of intellectual magic that can repel leaders of the reality of business management. Like children before powerful new toys, they forget that the model is only a mechanism simplification of reality.

The model simplifies life, to understand it better, but it does not eliminate it. Moreover, the model can forget assumptions that are underlying the analysis. Thus, the first big bet guess is the strategic segmentation, and so the current definition of strategic activities seems adequate and can produce a competitive advantage. Nothing is less certain, because the validity of segmentation is influenced by the actions of competitors and customer behavior, which are constantly changing. So vigilance at all times is required to ensure that, for a work on a sufficiently credible representation of reality, the assumptions are confirmed by reality elements what we have access.

Finally, all decisions are made based on projections of future behavior of markets and important players. These projections are often speculation, because the future is still very difficult to predict. Moreover, if we take into account that the political issues within the organization are significant, access to resources by the success or failure of a leader—a leader receives more resources to manage its business, more is considered a good manager—and the temptation to disguise or distort reality are also considerable. Let us take an example. Let us suppose that in a company the resources are granted only if an activity has the characteristics of profit or growth (or both) defined. Assume also that by slightly modifying the data, a strategic business leader can meet these requirements. As he knows that the uncertainty is large, he may decide to "cure the Don-born" to meet the requirements. This will gain him some time to keep intact his chances for personal success in the organization.

Thus, despite the usefulness of the analysis models mentioned the difficulties they cause have led to a take closer look at how companies are doing to cope with these difficulties. It was then noted the importance of the process by which the efforts of members of the organization are managed in complex situations. We had a close look at the process by which decisions were also taken. Decision-making in complex systems is so elusive that we cannot admit that it is incremental and therefore "less than perfect" when compared to the traditional rational model requirements. It has been proposed including three perspectives or models—rational models, organizational and political—to show how the decision meant not only logical analytical choices but also had to take into account problems of the operation of devices that were used to achieve the policies, and finally the preferences and actions of key individuals participating in decision-making. Finally, it was shown how, specifically, the rational models, organizational and policy combined to explain the decisions and actions in a situation of great complexity.

To simplify, we can say that decision-making in a large complex organization assumed a kind of "vertical specialization" of the management task as strategic action-related ideas can only come from people who are in touch with the realities of the environment and the organization. Their knowledge of the community and the environment allows them to make a strategic analysis. Their task is essentially strategic. The people at the top really have no way to assess the validity of the proposals made to them. They can neither redo the studies nor evaluate without devoting so much time and energy to this and by that time the entire organization can be paralyzed. They can then only say yes or no to what is proposed. They do this with the help of middle managers, who know better the ground realities, and better managers of the land requirements of the summit. These managers then make a task of translation and reconciliation between levels, a task whose characteristics are essentially interpersonal.

The people at the top, however, are the guardians of the rules. Not only can they identify key persons of "strategic" levels and "intermediate" mentioned above, but they can also change the structural arrangements, including rewards and punishments, so as to encourage desired behaviors, including behaviors of managers at the operational action. To do this, they need the help of middle managers, including their presence with strategic managers, and their understanding of cause and effect in management. Middle managers then have a decisive role to play in the system. How can such a set function properly? Why do not we have the same shortcomings as those that have been mentioned in the discussion of the portfolio model?

The first studies showed that the management of such a system requires monitoring and constant adjustments. The leaders at the summit should run their business by changing the rules of the game bit by bit, with the help of middle managers. They have the interest to cooperate better, because their credibility and their future in the organization depend on it.

Indeed, if the performance of a middle manager is good, that is to say, more advice and projects supported were wise, and there was more influence on the decision at the summit, the rewards will be great. To use a North American metaphor, it seems that, as a baseball player, he is judged on his "batting average." So try to choose the best projects and make the best decisions, from the perspective of the organization as a whole. Similarly, executives at the strategic level in their interest do propose, at the intermediate level, the decisions that will enhance their credibility with it. Again, the batting average is important. We do not generally seek to play games that are artificial to deceive the organization.

Thus, the system is constructed to reduce the risk of dysfunctional political games for the whole organization. Talented intermediate leaders are to precisely recognize the proposals that will strengthen the ability of the organization to survive long term. It is at their level that the management of the business portfolio can be achieved. The talent of the top leaders is to have a sufficient understanding of the functioning of the organization to build and constantly adapt the structures and rules of the game to produce the most favorable behavior for the survival of the organization.

The rational or strategic model dominates at the operational level, that of the definition of the decision. The organizational model, concerned about the nature and operation of devices, dominates the institutional level, that of context management. The political model, or interpersonal, dominates the intermediate level, the management of the pulse (which allows a decision to be brought to the attention of leaders who can say yes or no). Managing a complex organization appears as managing managers that allow the organization to make the best decisions for the survival of the overall organization in the long term. Thus, the complexity goes from the normal management of people and their actions (the management) to leaders involved in all phases of the decision.

Managing a complex organization is no longer about managing all decisions throughout the organization. It is mainly about managing the leaders who take decisions of a strategic nature, that is to say, the choice of areas and navigation in each of these areas, with the help of intermediaries' leaders to make this feasible task. Managing complex situation requires first and foremost that makes life easier for those who have the responsibility to lead the organization. In these cases, the operational action, that is to say the ability to integrate the operations of the entire organization, requires the development of decision-making models which enable better appreciation and better control of the relationship of cause and effect, and which better advise the members of the organization the directions that senior management values.

Managing complex situations means therefore to deal with cognitive limits imposed to the leaders and all those who contribute to "build" operational activities of the organization. There are two great ways to meet these limits: (1) by developing "content models" that suggest specific decisions—the portfolio model is one example; and (2) by developing "process models" descriptive of the operation of the system without suggesting the decisions, that reveal enough about the mechanisms by which the overall behavior is shaped, so that the leaders can design the necessary decisions.

Both models are useful, but higher the complexity, the greater the models feel simplistic and inadequate. It is then necessary to turn to the process models that are often more rudimentary. The great problem of content models lies in the fact that the behavior of the organization in complex situation is less linear and that there are more decisions that are generally enough to be applicable to situations anew. Experimentation is the crux of the complexity of situation. When examining the history of large complex organizations like GE and Canon, and the actions of their leaders, one is struck by the originality of their approach. One is also struck by their willingness to take head on the unique problems they face and the unique solutions they adopt. To facilitate experimentation and learning, we must focus on the construction of the system and its constant adaptation. That is what the process models do. They focus on understanding the functioning of the organization and what influences their operation. They provide the tools, without proposing a specific solution. From this point of view, they seem to be difficult to use with managers who are pressurized or frustrated by the hardships of daily life and experimentation managers, but they are unfortunately irreplaceable in complex situations. Process models show the importance of management to the success of organizations. They also reveal the importance of managers. If they do not want to play their role, because management is too taxing, it is surely time to replace them.

CHAPTER 16

Optimization of Business Diversification

Diversification is a common phenomenon in companies. Many reasons militate in its favor. Diversification occurs naturally in the context of growing a business. For instance, DuPont has become a big chemical company that we know today by gradually building on its core activities that were the manufacturing and marketing of explosives and, later on, by moving away from its core products. With the start of the First World War, the small explosives company had taken a considerable scale, and one of the first problems to be solved was the disposal of the by-products because there were no viable customers. This was termed as aromatic chemicals (benzene, toluene, etc.), that is to say products which were the foundations of what we today know as petrochemicals.

As the DuPont Company had a lot of additional resources, such as talented managers, scientists, highly experienced engineers, and surplus funds, it assigned its people to the evaluate these by-products, thereby engaging in manufacturing dyes, synthetic leather, nylon, etc., and thus, it began a process that would make this business into a large chemical industry company. In general, diversification is often stimulated by the presence of excess resources. The desire to use these resources effectively pushes the company in directions that are more or less different from those that exist. These balanced efforts in the use of resources are an important reason why companies are diversifying their activities. Diversification can also be stimulated by many other factors, such as the occurrence of unexpected opportunities or threats, the dynamics of competition, the pressure of the stock market or government pressure, etc.

Diversification is a measure of very common growth.

As companies inevitably reach the limits of growth due to the limitations of their industry, they should attempt to extend their activities also in other industries, that is either closely connected or sometimes unrelated manner or "conglomerate." But diversification raises the same issues of creating value growth in the home business industry. Diversification is justified only when there is value creation for partners. Does that mean creating value for partners? First, we must identify the relevant partners: usually one prominently includes shareholders, but we consider that the creation of value for shareholders is sustainable only if other key partners, such as employees, suppliers, or sometimes customers, find their accounts in the planned diversification, making the calculation value created for shareholders, who are calculating residual value, where the interests of all other key partners have been reasonably addressed.

Diversification can be through acquisition or internal development diversification, although internal development is carried out using the internal resources of the company as a basis for new activities. This kind of diversification requires an explicit strategy, dynamic and fruitful research and development activities, an organizational capacity to protect and develop new activities, and a lot of time.

Companies like 3M are typical companies that have been able to diversify through internal development. GE and DuPont were also precursors for a long time in this field. One of the few studies on this subject, conducted by Biggadike (1979), covering the practices of 40 major US companies, showed that the average time for a new activity to produce a positive rate of return of investment is eight years. Nowadays, it is likely that we can reduce this time, but is possible only when leaders find a formula to manage the complexity associated with the proliferation of activities. Japanese companies, Sony and Canon in particular, showed that it was possible. But, for many others, this exercise remains a cross. In contrast, diversification through mergers and acquisitions may take only a few months to be profitable. For this reason, diversification often raises the possibility of mergers and acquisitions.

Sometimes this type of diversification is being planned, as is often the case in large traditional companies, but this is not the case of acquisition. In acquisition, candidates present themselves without warning, the market conditions become favorable or unfavorable without warning, competing offers completely upset the logic of the acquisition and exert more pressure on policymakers. Therefore, taking the opportunity with determination can be viewed as essential, because diversification often raises the possibility of mergers and acquisitions.

Diversification through mergers and acquisitions can save time in a business or it can reduce the cost of entry into a new industry. Diversification by internal development cannot be the solution when the situation is stable, and there is no way in the replacement market, which provides quick access to the desired growth in the desired areas. Economists have often treated this subject, namely "make or buy" as an optimization problem in the use of resources. In this chapter, we rarely unlink mergers and acquisitions of diversification. Given the considerable activity of mergers and acquisitions in recent years, it is useful to take a fresh look at diversification in general, and diversification through mergers and acquisitions in particular.

Our goal is to help managers develop a program in this area that ultimately serves the interests of the company and its partners, including those of its shareholders. Our focus is primarily strategic; however, mergers and acquisitions are often approached from a purely financial point of view. We first mention the reasons that lead to diversification. Later, we will discuss the historical development of mergers and acquisitions as a phenomenon associated with diversification. Then we discuss the diversification strategies and the reasons behind diversification through mergers and acquisitions, including the need to create value.

Finally, why diversify? The word "diversification" is used to mean a lot. Ralph Cordiner, who was president of General Electric (GE) until 1963, spoke of "developmental diversification (RD)," "functional diversification" of "product diversification," "customer diversification" of "geographical diversification (international)" and diversification of funding. A restrictive definition limits this concept to the diversification of products and markets that we generally consider here. Many reasons underlying the decision to diversify: the need for growth; the need to balance the use of resources; the need to acquire new resources or maintain existing ones; competitive dynamics; the intervention of external powers (regulation, public policy, business controls, etc.). We describe here the 14 most common reasons, knowing that leaders are in regular news to justify their actions in the matter.

The most common reason is that a firm is constantly pushed toward growth and that its "growth driver" is predictable. It will grow by expanding its product or current market or it will go toward new products or markets.

These four possibilities can be considered as diversification, except, perhaps, the growth in current markets using current products, which could also be defined as a greater penetration into the sector in which the company already operates. However, as soon as one makes this penetration through mergers and acquisitions, such as the Royal Bank had done in acquiring Royal Trust, we find ourselves in markets shades, sometimes in product nuances that some might consider new, which would justify the talk of diversification. Ansoff also suggested that proper diversification, that is to say when moving toward new products and new markets, takes place in the following situations: when goals cannot be achieved by the expansion (market penetration, product development or market); when funds exceed the expansion needs; when the diversification benefit promises are larger than expansion; when the information does not compare, in confidence, between the profitability of diversification and the expansion.

Another reason for diversification, which is similar to the previous one, is the pressure exerted by lower growth in its original area. For example, Alcan began to diversify its activities in 1980 in the applications of aluminum because the aluminum demand began to show signs of slowing. Similarly, in the 1980s and 1990s, beverage companies like Coca-Cola or Pepsi-Cola began to enter into beverage markets other than its original products. In general, this situation applies to all companies that have significant business or even dominant in one area and consider that this activity has reached a stage of maturity.

Another pressure sometimes comes from research and development (R&D) efforts. Technological dynamism of a firm and the development of new products are important for diversification of a company. Thus, Sony is diversifying into electronic products; DuPont, in chemistry and its applications; 3M in consumer products; and most multinational pharmaceutical companies in new applications of their skills. In fact, all documentation was done to examine the process by which one can boost the creation of new products and, in doing so, moving "natural" of entre taken out of its current activities.

Some companies have advantages or underutilized resources such as the control of a distribution system, and wish them to be more profitable. This was the case when Gillette introduced in its distribution network blades, other consumer products, such as complementary products for shaving, lighters, pens, and even surfboards. Many oil companies have also developed networks of supermarkets attached to their gasoline distribution networks. These benefits can also consist of a management capacity out of the ordinary. General Electric believes that this advantage has made it successful because it led her to diversify into new high-tech industries and new services. That is why today, in some sectors, including GE Capital, the company is conducting multiple acquisitions as it is able to integrate and manage better than other companies. In the 1980s, the president of Daewoo also used the internal know-how in management reorganization as leverage to make acquisitions. The company has now become a powerful conglomerate.

The financial resource is often a cause of diversification. The availability of surplus funds to needs often led the company to consider diversification, including mergers and acquisitions. This happened to most businesses, including the recent unfortunate event of the French company Vivendi. Vivendi made from the company "LYEA," had, thanks to the latter, an important source of funds. This led her to consider diversification. The media was considered a chosen field, and thus gradually the company started in publishing, advertising, television, and, merged with universal music and cinema.

Sometimes, competitive pressure can lead to unexpected diversification. For example, if competitors go to unexpected areas, we tend to copy them. This is especially true with regard to geographic diversification, but we also find this in the diversification of products. So when BP began after the oil crisis, to diversify into other energy sources, it was gradually followed by Shell, Exxon, and all other multinationals. When the Continental Can Company began to diversify into products other than metal cans, all its immediate competitors had followed. Closer to home, in the late 1990s, Hydro-Q's diversification through acquisitions, particularly internationally, was also justified by the actions of major North American competitors.

The actions of competitors can also provoke retaliatory actions that explain diversification. Thus, in the 1970s, Xerox's interest in electric typewriters prompted IBM to consider making photocopiers. In consumer electronics, this type of warfare is constant; Sony, Matsushita, and others follow by systematically copying. The establishment of US multinationals in the auto market, before the globalization of it, was often motivated by the desire to keep distance companies competitors by putting himself in a favorable position to disrupt their main market.

The globalization of markets and industries has created a new competitive dynamic where economies of scale and scope are becoming more spacious. Many major acquisitions, particularly in the automotive, telecommunications, and the Internet, and at the regional level for many other sectors (railways, printing, food distribution, etc.) seem to have had justification for the emergence of new market logic and encompassing larger spaces.

Globalization is often accompanied by a surprising convergence in many areas previously considered relatively tight. Thus, in telephony, computer software was integrated into the conventional switches. This generated so many risks for companies in both sectors as we saw cross acquisitions of major companies in the two sectors. Thus, Microsoft acquired a telephone company, and several telephone companies positioned themselves in the computer software market.

Some companies are diversifying to avoid takeovers. This happens mostly to companies that have unused resources, poorly valued by the stock market, and that could be the envy of "financial sharks" looking for opportunities to make money quickly.

It also happens that personnel issues are so important that they justify diversification. In particular, when one wants to attract or retain top talent, they can be encouraged to go into new and attractive areas. The Japanese companies have experienced this situation during the recession of the Japanese economy in the 1980s. Many spin-offs (spin-off), sometimes stimulated by surplus managers were supported by large companies, as there have been not only opportunities to keep near them valuable cadres but also interesting opportunities. In general, many companies think that to attract and retain world-class managers, we must offer them opportunities for development that allows diversification.

The desire to reduce the effects of the economic cycle and changes in net cash flow that can result prompted many companies to design business portfolios balanced in the matter, which led them to areas of new business. Bombardier has deliberately built a portfolio of activities that allowed it to balance inflows and outflows going toward different segments of the aerospace manufacturing, to different segments of transit, and to the financing.

Government actions, including deregulation, can also stimulate diversification. Thus, in the United States and Canada, changes in banking laws led all banks to diversify into a range of financial activities (portfolio management, national and international investment, brokerage, insurance, financial advice), on the sidelines of the activities of deposits and loans. The same can be said of the activities of transport companies (rail and air), which have become more international in scope (and sometimes much more diversified in terms of products).   Canadian National is now a transportation company that covers all North America. Similarly, the train or RATP30, specialized French company in transportation, provides consultation and selling, everywhere, on know-how and technology, as would a consulting firm in the field. In newly industrialized countries, the government's economic policy has also been an important factor in business diversification.

Korean chaebol are the product of government policies of South Korea. In some countries we have seen deposit boxes and investment. Finance and investment companies make acquisitions for reasons that are probably more related to socioeconomic development than the traditional business logic. The internationalization of Chinese enterprises, especially in building construction and energy, follows the same logic.

In the United States, antitrust laws have also pushed many companies that were in low-growth areas to seek growth through acquisitions outside their area of origin. Thus, Exxon tried, with little success, to diversify by going to the office systems market. The chemical entre taken also tried to get out of their traditional markets. This resulted in major interpenetrating fields of biotechnology, food processing, and pharmaceuticals. Finally, Microsoft gained control of full repositioning connected with major diversifications.

All these reasons justify the diversification of logical operations, but do not overlook some emotional reasons, such as preferences of leaders and their desire to build great empires, to leave their mark in the industry, the thrill of the bet, etc. These emotional reasons may accompany the reasons mentioned above, but they can also play a key role. Governments also intervene according to a logic that is not that of the business world, where national or merely partisan interests are involved. The efforts of the French government to marry companies Gaz de France and Suez is based on the same logic.

What if we make a history of mergers and acquisitions to better understand this problem? Mergers and acquisitions are a phenomenon almost as old as businesses themselves. However, systematic surveys of mergers and acquisitions have started in the US at the end of the 19th century; in Canada, the data is a little more recent. Salter and Weinhold (1979) demonstrated that this practice in the United States has manifested cyclically. We have completed the work of these authors in formulating the proposition that there has been, since 1976, a fourth wave (Christiansen, 1987) which shows that mergers and acquisitions remain at a high level.

The first wave of mergers and acquisitions lasted from 1895 to 1904, experiencing a peak in 1900. This wave was characterized by Stigler (1950) as "monopolistic mergers." Indeed, there have emerged many of the large companies today, especially the descendants of Standard Oil of New Jersey, US Steel, GE, United Fruit, Eastman Kodak, American Can, American Tobacco, US Rubber, DuPont, PPG, International Harvester, etc. This wave has been driven by the development of the railways, which opened for the first time the US market as a single market. Economies of scale were like today globalization, the dominant factor in deciding the conduct of a merger or acquisition. The end of this wave occurred around 1903–1904 and coincided with a major economic recession. The second wave, which occurred in the 1920s, began around 1922 and ended in 1929 with the famous stock market crash and global depression that followed, from 1930 to 1933.

Stigler described the wave as "oligopolistic mergers." Indeed, this wave has given rise to strong "number twos", increasing the degree of concentration in most manufacturing industries. Thus has emerged the Bethlehem Steel, Allied Chemical, Continental Can, etc. During this wave have also emerged major holding companies in the fields of manufacturing and distribution of electricity, gas, and water. This wave has been driven by the development of highways and automobile, that offered a promising and viable alternative to rail transport. The third wave began after the Second World War and lasted till the late 1960s; however, this wave did not hit the big traditional businesses. The acquiring companies were generally small or medium sized and were often remote acquisitions of their home areas. This has led to companies of a particular type called conglomerates, with operations in unrelated fields.

These acquisitions were clearly associated with the diversification. This wave has been driven by high technology, and a lot of new companies who saw the day wore brand.

Thus prospered the likes of Litton companies, Raytheon, Teledyne, Textron, United Technologies, and Trilon Genstar. Speculation had reached record highs. Malkiel called this "the Tronics Boom" because many companies having no value can be easily sold on the market as long as they had "Tronics" in their name. In 1973, North America experienced the most severe recession since the depression of the 1930s. Since 1976, there has been the emergence of a fourth wave marked by different strategies on the part of firms involved. According to Christiansen, they are of three types: growth, focus, and create barriers. Under the influence of a need for growth (the first type of strategy), there has been consolidation of new fragmented industries (software, health food, etc.) or the application of new technologies to older industries (factory automation, computer use in entre taken as Xerox, for example). The focus (the second type of strategy) aimed at increasing margins and efficiency (Renault, America has tried to do that to reach a market size required in the small car sector) to recover an identity and a position in the market when the diversification of the previous period had led to diversion (e.g., GE), or to ensure stability and performance.

The third type of strategy (creating banners) sparked conventional vertical and horizontal integration, which have achieved economies of scale or scope or, again, to restore technological leadership that had collapsed (e.g., DuPont). We can say this wave as "strategic." It is dominated by the search for a favorable strategic position and the quasi-revolutionary influence of information technology in all industries. The movement of mergers and acquisitions declined in the late 1980s, but it remained relatively high. In the 1990s and 2000s, a new outbreak of mergers and acquisitions that we tend to relate to globalization and the revolutionary effect of the convergence of computing and communications technologies was known. This new wave seemed to bring us back to a combination of the desire to achieve economies of scale due to the globalization of many industries, the desire to refocus on global segments, the desire to protect major turbulence in some industries and speculation whose dimensions exceeded those of the 1960s; it is still too early to characterize this new wave, but we can already say that the number of transactions and the amounts involved were considerably higher than in the preceding waves.

To a great degree, the M&A market at the dawn of the 21st century represents a synthesis of the myriad forces impacting business and the expeditious way managements are dealing with them. The fallout should include more combinations of giant concurrents, escalating price thresholds for mega-transactions, increasing incidences of hostile bids and bids contested, an obsession among corporate survivors to seek first-mover advantages and concurrents to beat to the punch especially for the most desirable targets, and, perhaps, less worries about niceties like pure strategy synergies. It is an overactive decade. The race for the most favorable position is accompanied by movements, almost like Napoleonic military, instead of reflection and strategic integrations.

Diversification is itself a business strategy. It is a decision that leads a company beyond its current activities. This expansion of activities can be so close and compatible with current activities or be in completely different areas, not related to current operations. As we have seen in our historical overview, the waves of mergers and acquisitions have in fact been dominated by an underlying strategy of diversification.

The first wave, whose zenith was located in 1900, was led by market control strategy, with a focus on economies of scale and buying competitors. The second wave, that of 1920, was made possible thanks to antitrust decisions in the United States. It looked like the first, with a focus on the acquisition of smaller competitors to create companies able to compete with what remained of the large companies born from the first wave. The third wave, in the 1960s, has been a wave of conglomerate diversification outside the territories in which the acquiring firms were. Then, in terms of the waves of the 1980s and 1990s, the strategy was more explicit and more diverse, bringing all the possibilities of monopolistic temptation.

Rumelt (1991) constructed a typology widely used today on the strategies of companies in general, and that applies particularly well to strategies for diversification through mergers and acquisitions. He proposed four major types of businesses according to the chosen strategy:

(1) Enterprises' simple activity (single-business company) which, as the name suggests, are companies with a strategy of focusing on one activity in one area of activity.

(2) Dominant business activity (dominant-business company), in which the major field of activity (single activity or vertically integrated activities) represents 70–95% of sales. General Motors, Texaco, IBM, Scott Paper, and Alcan were typical businesses in that class until the early 1980s.

(3) Businesses-related activities (related-business company), which are diversified by adding activities that are connected in a tangible way to their strengths and their expertise. In this case, no activity accounts for over 70% of company sales. DuPont companies, GE, and General Foods in the United States, Bombardier in Canada, and BSN or Rhone-Poulenc in France, at the beginning of the century, are typical of this class.

(4) Companies' unrelated activities (unrelated-business company), also called conglomerates, which have diversified without worrying

links between activities. No activities will account for over 70% of sales. Among the representative companies of conglomerates are the chaebol in Korea, Chinese big family companies, or Taiwanese Onex Canada, and Vivendi in France.

This categorization is generally well accepted everywhere and in all disciplines. The links between the new business and core activities are an essential element in defining the strategy of diversification. These links also appear in the work of Rumelt (1977), as a tool permit that reliably predicts business performance. Thus, companies whose diversification is connected seem to eventually achieve better performance in terms of profitability, with conglomerates having the least good profitability. By cons, as you would expect, conglomerates arouse the greatest sales growth and stock price. These works were carried out on a sample of the wave of diversification through mergers and acquisitions in the 1960s, and all the work that has been done since tend to confirm these results. We provide the support for the curvilinear model; that is, performance increases as firms shift from single-business strategies related to diversification, and performance decreases as firms' aim changes from diversification related to unrelated diversification. These results are not surprising. The concept of strategy suggests that the company should not move away from what it does best. As a result, the most interesting is the strategy of the related diversification. It generates synergy, that is, real opportunities for value creation. It assumes that the company is trying to find and make a link between the new activities and market products in place.

If management skills are critical elements for success, then it is appropriate to classify diversification strategies connected using these capabilities as criteria, as we offer further. In contrast, a conglomerate normally does not seek to exploit existing capabilities. He expects little transfer of knowledge between activities. Companies that diversify so connected can be divided into two groups. There are those who go to the market, functional products that require similar skills to those available, such as using the same distribution networks, the same facilities or even the know-how of production, the same marketing skills, etc. This strategy of diversification can be called additional connected.

The second group of companies is represented by those who are diversifying by adding expertise and functional activities (marketing, distribution, production, etc.) to their current base. This strategy can be called connected complement. The pure form of this strategy is vertical integration, as when a producer of crude oil purchases refineries and gas stations. The horizontal axis measures the addition of market products, while the vertical axis measures the addition of functional activities. Mergers and acquisitions can of course involve both addition of functional activities and extension for contracts-products, but if the ruling is functional, we talk about complementary connected diversification, and if the dominant key is market-products, we speak of additional connected diversification. An important question is to ask is whether the diversification strategy creates value.

Creating value involves determining for whom value is created. Without entering into the debate on the importance of stakeholders, we make a heuristic assumption that the ultimate target group to meet once every other group taken into account is represented by shareholders. We talk about creating value for shareholders. Value creation can come from any of the strategies outlined by Rumelt. It involves one of the two following options, or both: (1) increase yields, that is to say, income streams, for diverse business (after merger or acquisition) beyond what may be achieved by separate companies (before merger or acquisition); (2) reduce the risk of diversified business below the risk level to the companies before the acquisition or merger. This is what we will look at now.

It must be said that a diversification strategy makes sense only if it helps generate value for the company's shareholders. Value creation in particular assumes a favorable positioning of the company compared to its competitors and has to maintain this advantage in a sustainable way. Sustained favorable positioning is generally based (Andrews, 1987; Prahalad and Hamel, 1990) on quality resources and a balanced and dynamic inner workings. This should result in a favorable risk-return profile. Salter and Weinhold (1979) proposed a systematic approach to dealing with the creation of value. They suggest that reducing risk and accruement returns, beyond which it enables simple financial diversification portfolio, can be done through the following actions.

In terms of increased yields, we must say that for companies that are diversifying by mergers and acquisitions, there are six different ways (see following pages) to generate returns that exceed those obtained by an investor diversify his portfolio shares. The first three methods are more relevant to a related diversification, and the other three are more for an unrelated or conglomerate diversification. To better appreciate the benefits of diversification, connected or not, it is useful to use models that affect the operation of entre taken and their evaluation by investors, including policy templates, portfolio of products and markets, and financial resources. These models suggest that the more acquisition is related to expertise and resources of the buyer, greater are the potential profits for shareholders of the combined company. In particular, the concept of strategy suggests that the benefits are only generated when there are resource transfer opportunities or skills between partners of a merger or acquisition.

This is what normally increases the productivity of an investment in the combined company and therefore creates value for shareholders. This creative transfer value is what is usually called "synergy." Concretely, this manifests itself as follows.

The M&A diversification can increase the productivity of capital when the particular skills and detailed knowledge of the industry of one of the partners can be used to strengthen the resources and capabilities of the other partner (and therefore, help to better take advantage of opportunities or to better deal with threats that come to him). Thus, the traditional skills of Bombardier for government relations have been very helpful to the development of Canadair and later, to all acquisitions and aeronautical activities of the company. Similarly, Royal Bank hoped at the time of the acquisition of Royal Trust, crossword benefits in terms of customer relationships and sharing knowledge In France, the successful merger of BNP and Paribas in 1999 was based on the same hopes.The investment in markets that are similar or related to current activities can lead to a reduction in long-term average costs. These reductions can come from effects of scale and rationalization of production costs and other managerial activities. Thus, the acquisition of IG by Transcontinental Group in 1998 was to put rationalization in marketing, human resources management, financial management, supply management, and generate economies of substantial scale, reducing the long-term average costs very significantly. We can say the same of the BNP Paribas merger or acquisition of Canada Trust by Toronto-Dominion Bank.

The expansion in the areas of competence can also generate a critical mass of resources to do better than competitors. When a small business makes an acquisition or merger in the near field, it can have access to additional resources (money, staff talent, skills, etc.) which can then be used to develop skills at or above those of the established competitors. All the mergers and acquisitions of "the wave of oligopolies," mentioned earlier, were intended to do that. The tremendous development of GE Capital since 1980 enabled this, before the armada of traditional financial institutions.

One of the most frequently mentioned arguments to talk about the benefits of unrelated acquisitions is the stabilization of financial flows. The argument suggests that the diversified company can work in activities against cyclical, which allows a business to be on top when the other is hollow with a relatively stable average. This argument is obviously superficial. First, it is difficult to find activities perfectly against cyclical, but, more importantly, the benefits of stabilizing the flow can be obtained by the single investor when using the instruments available on the financial markets. Even if the direct benefits are overstated, some indirect benefits would be substantial, as the capital increases efficiency, in particular, through the centralized liquidity management and the management of debt and debt capacity.

Another direct benefit is the possibility of development of a certain cash flow, in particular for high growth companies or capital intensive. This would avoid double taxation (corporate and personal) of the individual investor, while achieving the same results, if the decisions of the company and the individual investor are identical or similar. Somehow, by doing this, "we buy cash" with certain acquisitions. Of course, be careful to the actual cost of liquidity and their actual availability in time. In connection with this idea of buying liquidity, there is the idea, more contemporary, to use surplus cash to maturity activities to support new and promising activities. These arguments can be summarized as follows.

The diversified company can operate as a bank that takes the funds generated by units with a surplus and then directs them to those in deficit, reducing the need to resort to the financial market for working capital needs. This benefit is intended operational (with reduction of transaction costs) and has nothing to do with the allocation of resources for investment, as described them in the presentation of the product portfolio model.

Diversified companies may also, under the portfolio model, use the resources generated by the units that have net cash flow (cash flow) to provide high investment funds to units that currently have negative cash flows or zero, but the prospects are promising. This can improve long-term profitability of the entire company. This is the same principle of the product portfolio model. This is a factor even more decisive than the diversified company, since it acts internally as a market in which it would have inside information not available to the investor in the financial market. It has access not only to a greater number of investment opportunities but also to a better appreciation of the costs and benefits that would normally allow it to make better investment choices. The allocation of corporate resources process is then more "efficient," that it cannot be found in the market.

By aggregating the risks of its various activities, the diversified company can also reduce and achieve a lower cost of debt than can separate activities. This also allows it to have more leverage. In general, the cost of capital decreases and yields increase.

The reduction of risk diversification is closely related to increasing yields. Indeed, the risk is often an expression of the variability of cash flow generated. Anything that reduces this variability is seen as a risk reduction. Thus, in particular, aggregation of risks (risk pooling) allows for this by reducing both the liquidity requirements and changes in flow. The total risk of the diversified businesses can thus seem lower. But it is not always the case.

Indeed, the perception of professional investors risk is also related to the quality and transparency of information available on a company. The highly diversified businesses can also be perceived as more risky because they are less transparent. Furthermore, having more debt, which is a potential advantage for the company diversified, can also be considered by investors as more financial risk. Anyway, it is assumed that the diversified company is often able to provide more stable flow that can help an investor by diversifying its investment portfolio, thanks to six main actions that we talked about earlier.

Diversification by M&A creates economic value if the present value of expected returns is greater than the cost of the acquisition. When the price of an asset changes, this reflects the need for a reassessment made by market participants, the size of life, and lead-time of future net cash flows. It may also reflect a greater ability to predict the characteristics of the flow in question. If predictability increases (if the variability of flow decreases), then the risk decreases and the value of the underlying assets increases. Managers know the value of their assets increases when they reduce uncertainty, perceived by the market in terms of expected returns. Salter and Weinhold (1979) suggest that in assessing the creation of valor, you have to use three models: the strategic model, the model of carrier sheet, and the risk-return or financial model.

The use of these three models is described by Alain Noël (1987). Each of these models reflects the need to address the risk and uncertainty of business, but each model approaches this need differently. The strategic model focuses mainly on business activity or single units of multi-product and multi-market companies. This type can be called "operational." The risk faced is of two types:

- In terms of management judgment, on the variability of returns of their business: This judgment is exercised by answering questions like what factors can negatively impact this business and what are the chances for them to do?
- In terms of their ability to assess the future financial performance of the business through the process of budgeting and resource allocation: This method is essentially a forecasting method based on the relationship between the budget and actual results. This will find the activities whose performance is the easiest to predict.

The portfolio model focuses on the management of the product portfolio and enterprise markets. This management is made at the corporate level, undertaken by a group of value of the portfolio (VP). Obviously, it is also interested in flow of funds, but for the entire portfolio. We try to reduce risk by stabilizing the flow. We then tend to choose a portfolio that includes activities at different phases of their life cycle, so there are more mature activities to fund emerging business, both for their daily functioning (i.e., working capital) and their development (that is to say, the investments required to achieve a dominant market share).

Doing so reduces the negative perception of the investor in respect of the risk. The financial model or risk-return adopts the perspective of a rational investor, knowledgeable and acting in a reasonably efficient market. The type of analysis is that of the financial market. Risk measures have been set by sophisticated statistics. But fundamentally, we measure the volatility of returns of a security to assess the risks. As investors can diversify the specific risk to a title by building a portfolio equivalent to the market, analysts consider that the only relevant risk is "market risk" or, if preferred, the part of the risk of a title that is related to the overall market. In financial jargon, this is called systematic risk or market risk.

From these risk assessments, we can talk about the value of a business as its market value. This is the best way to appreciate the worth of an asset or revenue stream. The market value is what an informed and motivated buyer is willing to pay a vendor for a title or an asset. Buyers and sellers confront their opinions on what the security or asset will generate as revenue streams and agree on a price that represents a compromise between those views. The finance studies show, and generally admits, that the market value of a financial asset can be calculated using the standard formula of discounting. Of course, the investment or, in this case, the M&A should only be made if the market value, i.e. the expected value, exceeds the cost of the investment or the M&A.

These three models are complementary. The risk-return or financial model suggests paying attention to the relationship between the returns of an asset and those of the economy as a whole. It suggests that managers must develop strategies whose aim is to have better control of cash flows (cash flow) and future returns on their assets. However, the policy templates and portfolio of market products provide the manager a methodology to do this.

Leaders who choose to diversify their activities must first decide if they want to do so connected or not connected, keeping in mind the benchmarks that have been provided throughout this chapter. They should especially remember that related diversification, especially through mergers and acquisitions, should allow some synergy in the use of resources in both partners. However, the definition of what is the source of links is creative, and, in a sense, it is a highly strategic decision. People tend to consider as connected implying similar products or markets, technologies or similar scientific research or activities that complement the length of the same trade chain.

Diversifying conglomerate seeks no such links. But management challenges, particularly on strategic and financial plans, and various activities such as conglomerate diversification have a chance of success if the company has recognized those challenges and has surplus management talent. We can then say that the appropriate methodology for making related acquisitions is different from that which is appropriate for unrelated acquisitions. The first concerns with the creation of value that assumes that the individual expertise of the partners can be applied to the problems and possibilities of the other. While in the second, one especially seeks to improve the risk-return profile of all by making a more efficient management of funds or assets. However, it is important to use this grid, inspired by Salter and Weinhold (1979), taking into account the particular characteristics of each entre taken. Thus, a cash-rich company that expects steady cash flow over the next five years cannot have the same goals (therefore cannot fulfill the same way the evaluation grid) as a company that lacks liquidity and expects significant demands on the matter since its start of activities.

CHAPTER 17

Optimizing Global Business

The phenomenon of globalization of markets is not new. For modern enterprises, internationalization of activities is a natural stage of development (Chandler, 1977, 1990; Wells, 1976). In the past (Chandler, 1962), a company like Exxon (formerly Standard Oil of New Jersey) could consider the oil market other than as global. The strategy of this company, like its competitors, was thus built on a vision of the world which looked strangely like the current idea of the global village. The market that most companies think is a huge market, even when they deliberately choose to focus on only part of it.

Globalization thus corresponds more to the will of a particular company to go elsewhere, but it stems from the desire of all companies to do so. Even when a company does not plan to internationalize its activities, it should still expect that others from elsewhere come to defy its own territory. So globalization, which was a specific strategy a company has become a structural element that changed the nature of the competitive dynamics for all companies. The competition caused by globalization, as competition in a particular territory, poses significant problems for governments.

When competition is global, the overall logic is no longer under the direct control of a single government. It affects several governments at a time, and each one is tempted to act without regard to the interests of others.

Unfortunately, act unilaterally, without regard to the overall logic of an industry, can result in adverse outcomes for the country. Moreover, if the government is accepting the economic logic of globalization, he abandons the same time its regulator prerogatives of national socioeconomic life. This assumes that a government should try to understand and to influence the logic of firms in the interests of the nation, hence the idea of national competitiveness. In this chapter, we try to clarify all those ideas that become commonplace for the manager. The chapter consists of three main parts. The first is devoted to the dynamics of globalization. We first see what globalization. We then consider a set of specific questions about globalization and conclude on an overall reflection on the evolution to a global village. In the second part, we will discuss the global strategies and in the third part, we discuss the management of a global company.

The idea of globalization covers different realities. Thus, we speak of globalization of markets as if it were a reality independent of the will of the shareholders (Porter, 1986). The electronics market or that of the pianos is often presented as a global market. We talk about globalization of the company and its strategy (Doz, 1986; Porter, 1986). IBM or Ford are presented as companies worldwide. For a long time, economists are studying the phenomenon of internationalization of activities, especially trade, linking it to the fundamental economic paradigm (Caves and Jones, 1981). Ricardo's theory is both the simplest and the oldest. For him, shopping patterns were dictated by supply. So, a country should export food and clothing import if food production in terms of productivity was relatively higher than that of others. In this case, concerned countries should completely specialize in activities that benefited them. The model of Ricardo was the base from which refinements have been built. In particular, the Swedish economists Heckscher (1919) and Ohlin (1933) argued that countries exporting commodities whose production requires relatively intensive use of factors (labor, capital, raw materials, etc.), locally more abundant while trade in commodities tends to eliminate international differences in factor remuneration.

These major theories and all the sophistications that were made to them (Caves and Jones, 1981) do not explain the internationalization of firms, although they provide interesting insights into the internationalization of trade. We must move toward broader and more managerial theories for answers that are close to our concerns. The largest study on multinationals conducted at Harvard in the 1970s led to a particular theory of the internationalization of the company explains how and why major US companies have become multinationals. According to Vernon and Wells (1976), data on the internationalization of the activities of these companies show that it has completed a cycle linked to the life cycle of the product, with three phases: export, overseas production, and import. When the product has been launched, usually as the result of an innovation, the company has a sort of monopoly on it and then you just have to sell it on as many markets as possible. This is then the easiest and most obvious step. This will export to more or less distant countries, depending on experience and the capacity of the firm.

When the product reaches the beginning of the mature phase, the technology is usually already available for some competitors in the export countries. They then begin competing activities. Quickly, the competition becomes too strong and disadvantages of the location, too important. It is then necessary to establish in the markets in question, to enjoy similar production conditions. Finally, when mature or even declining, it may happen that the comparative advantages of some countries make it cheaper to import the products in the country rather than producing them locally. Thus, the product cycle generates a process of internationalization of the firm that is predictable.

In more recent work, particularly on the management of multinational companies (Doz, 1986; Porter, 1986; Prahalad and Doz, 1987; Bartlett, Goshal and Doz, 1991, 1992; Rugman, 1990), the "globalisation34" appears as the result of changes in the environment, including the structure of the industry and government action on the one hand, and corporate action on the other. Thus, in the automotive industry, considered by Doz (1986) as an industry in the world, the path toward globalization was caused or facilitated by three factors: (1) the formation of the European Community, which significantly increased trade between European countries; (2) the efforts and success of penetration of Japanese with substantial benefits in terms of productivity and quality; (3) reducing differences in consumer preferences in the US, European, and Japanese markets. The energy crisis of 1973 and the living standards of leveling have facilitated the evolution toward advanced technology and similar products.

Added to this are some fundamental characteristics of the automotive industry: economies of scale in production, distribution, and marketing; productivity differences between countries at the international level and the move toward free trade, which tends to push toward market consolidation or to link between them.

For a long time, governments have resisted the trend toward the globalization of markets such as automotive, but the movement was so strong that some of them decided to play the game more open, provided that social situations, including employment, are taken into account by the affected companies. Other governments have been obliged to follow, continuing to try to impose rules to safeguard national economic interests. The result is not a complete globalization, but a "managed" globalization, like any national market is "administered" provided that social situations, including employment, are taken into account by the affected companies.

Globalization is also the result of corporate action. Porter (1986) proposed to use the value chain as an element of appreciation. Companies are, according to him, constantly trying to determine where best to locate the various elements of the value chain. It can then occur that one or the other or more elements of the value chain are located in different countries, to take advantage of factor productivity differences in those countries. This location is what he calls a configuration.

However, the configuration options are constrained by the difficulties of coordination that too much dispersion could pose. We must then make a strategic choice to find the best combination of configuration and coordination. This combination can then be called a "global approach." Doz (1986) goes a step further and note that, depending on the nature of the industry, three major strategies seem to be frequently used: the integration strategy, national strategy, and sensitivity of multifocal strategy, a mixture of the two previous. When the industry is truly global, with few external constraints, the strategy adopted by companies is an overall integration strategy.

This is the case in the automotive industry where, subject to national constraints considered minimal, companies locate their activities in obeying only the law of the economic optimum. The only constraint on the integration strategy is a physical coordination constraint flows. When the industry is controlled by the government, as is the case at present in the space industry, the most appropriate strategy is a national strategy sensitivity, in which each national party company behaves like a small independent company, sensitive to national competitive dynamics.

Finally, when the industry is mixed, that is to say subject to both a structural dynamic that pushes for integration, and significant government intervention, as is usually the case of computers and microelectronics in Europe, the most appropriate strategy seems to be one that tries to combine the most favorable aspects of integration and national sensitivity. The appropriate combination can change from one industry to another and may vary from one region to another or from one company to another. There are at least three concepts, which partly overlap: (1) globalization of national markets; (2) globalization of the industry or competition; and (3) globalization of the company.

Globalization of a national market is directly related to lower prices. A national market can be said overall when the opening of this market to the presence of international competitors is sufficient that there is no reason that, for all important industries in the country, most of the large global competitors are active there and compete with each other. The globalization of the industry or competition occurs when the competitive situation in a country is linked to that of many other countries. This usually occurs as a result of the action of one or more companies trying to link their activities from one country to another to take advantage of a global competitive advantage. In this case, the only way for competitors answer is to find similar solutions. Globalization is also facilitated by a series of factors, including the standardization of education, access to technology, general access to information, and above all, the convergence of tastes and needs of consumers (Cvar, 1986). Again, one must think of degrees in the globalization of industry and competition. Finally, the globalization of the business is related to the collapse of its value chain. More activities of the value chain are scattered around the world, the more the company is global convergence of tastes and needs of consumers (Cvar, 1986).

The strategies of companies in globalization conditions are different strategies. Yet, the globalization of industries and reduced competition, perhaps temporarily, is the relative size of each player. Therefore, it makes the strategies of domination by costs rather random and difficult to maintain. This is why the differentiation and focus strategies are more frequent and often more defensible. The strategy of differentiation, in particular, is quite favored by globalization because it is often driven by the same factors. For example, the standardization of tastes created in the same market segments that cater to similar populations throughout the major markets of the world (e.g., automotive, electronics, clothing, cosmetics).

Globalization also creates many opportunities concentration. Thus, the Canadian company Peerless Clothing has focused on the production of high quality men's suits for the US market. Doz (1986) reminds us that globalization does not eliminate the government and its actions. In fact, there is no completely free global competition. For smaller players, it does not matter, but for the big players, this requires finesse in assessing the situation and the formulation of the strategy.

Involving the government to reflection enriches the possibilities. Considering the degree of globalization may be low, medium, or high, according to the discussion in the previous section, and if you consider that the government is interventionist or liberal, previously mentioned strategies become more or less relevant. So:

- When the government is allowed to make and that competition is local, the main shareholders find themselves in opportunistic situations. They try to leverage their local position to strengthen their overall position, but they must do so with caution not to trigger reactions that would increase global competition.
- When the government is allowed to make while the competition is in the process of globalizing, the most common strategies are those that take advantage of the competitive dynamic toward more openness. Doz (1986) mentioned including the integration or the location of production facilities so as to take maximum advantage of economies of scale. While globalization is not complete, integration must be cautious, because it leads to rigidity and very important commitments that cannot be easily undone. Probably the most flexible strategy in this case is differentiation. An interesting example is that of Becton Dickinson, the leader in disposable syringes.
- When the government comes shortly and that the globalization of competition is high, there is an accentuation of trends outlined above. Reduce costs can take a regional or global. On all possible segments, can also be limited to specific segments when these segments are tight enough on the cost side and image to customers. The example often cited, the latter two types of strategic location, is the automotive industry. In this industry, the economic scales for many important components, such as engines (more than two million units) and gearboxes (over two million) are so great that no national market is sufficient for existing production capacity. In some cases, however, as in some luxury segments where differentiation is the dominant strategy, companies generally produce lies in one place and serve a global market. The German and Japanese companies have used this strategy in the years 1980. The increasing globalization of the automotive market leads to combination of integration and differentiation sounds specific to each company.
- When the government can exert significant pressure on companies to take into account local needs, the necessary strategy is called "national sensibility." This occurs in industries where the stakes

for the country are considered important, while international competition is weak, companies with behaviors of mutual accommodation. This is particularly the case breweries in China since the early 21st century.

- When the government exerts pressure on businesses, but the competition is feverish, in the process of globalization, as was the case in the computer equipment industry or telecommunications equipment to the mid 1980s, type of strategy to be drawn is often a strategy of concentration or a mixed strategy of differentiation and integration. Doz (1986) described the strategic behavior multifocal (mixed strategy) of the telecommunications industry in Europe. Mixed strategies include alliances on particular aspects, such as the production of engines in the automotive industry (agreement between Renault and Volvo, for example) or the conduct of research in common
- When the stakes are important for governments on the one hand, and the competition is global and strong, on the other hand, possible strategies are either sway strategies, so by necessity mixed with alliances of all kinds, either strategies whose ambitions are to reduce the battlefield by the concentration. In the aviation industry, the two main players in the world (Airbus and Boeing) are, in fact, increasingly, business combinations with more or less permanent alliances.

Also, the globalization of markets, industries, and companies has significantly reduced the ability of governments to act directly to influence corporate behavior. They are then obliged to act indirectly by creating conditions that lead to the decisions taken Entre favorable government policies (e.g., job creation, local technological development, etc.). In doing so, the companies become customers of States, and competition among nations for their "favors" is exacerbated. The competitive ability of nations becomes a useful concept to consider. To measure the competitiveness of nations, Porter (1990) offers industry by industry make and use a simple model: the diamond of national competitiveness. The diamond defines the competitiveness of a nation in a given industry, such as its ability to encourage companies to make the country a platform for action in their international competition. The four constituent diamonds are:

- The characteristics of demand for industry products. Porter suggests more demand is demanding and sophisticated, most businesses should be drawn, as it should encourage them to develop competitive capabilities (products and technology) that would keep them in the forefront.
- The characteristics of factors of production, including labor, capital, and technology. More labor is quality, more capital is available at a cost and competitive conditions more technology in the country is considered advanced for the industrial sector, and the country is attractive to large companies sector.
- The characteristics of the structure of the sector and therefore competition in the relevant industrial sector. Thus, contrary to what one might think, a dynamic industrial sector, where competition is strong, without being wild and fierce, is preferred by larger companies because it maintains a healthy tension, the source of health and the strength of committed companies.
- The quality of support industries. A country attractive to a company in a given industry is a country in which there are complementary and support industries that are dynamic and innovative. When this is the case, we can expect that increase ability to innovate the whole industrial system, the ability to meet the needs of industry and the synergy between the system components. Note that the idea of the diamond suggests that a country without a strong diamond's interest to give up in the relevant industrial sector.

The approach of the Porter diamond is valid for companies or countries that are in a dominant position in a particular industry, but businesses in peripheral or marginal position are then forced to conceive the world differently. For these companies, it is better to speak of a kind of virtual diamond, the points can be dispersed all over the world in search of competitive advantage for the firm. Thus, it is possible that a local clothing firm believes that his country is advantageous for certain factors (capital, technology, and labor, for example), that the application is regional, and the competitive dynamics and industry support are international. In a way, the virtual diamond is a construction firm rather than a feature of the country.

With further the logic of the virtual diamond, one might even say that strategists at the national or regional level can adopt a pragmatic approach by examining the international conditions and the areas in which the companies of the country can play a role, even if it is not dominant. If the garment is in Canada, from this point of view, typical. It is clear that Canada has great benefits in regard to the availability and cost of capital. Similarly, all design technology and competitive workforce is very high quality. Moreover, it would be absurd not to recognize that, for supporting industries, especially equipment manufacturers, it is best to turn to the major international producers are the German and Swiss companies, in particular. Similarly, the US market, accessible and powerful, remains a reference.

The idea of national diamond is an interesting and useful idea for the analysis of the global competitive situation. The formal diamond should not, however, be considered as a special case of the idea, more generally, to virtual diamond. In this case then, we return to the idea of a diamond that would build a firm or group of firms, the state is trying to understand these diamonds to better influence them, so to promote the goals of which he is the guardian.

In terms of economic confrontation between nations, the substantial opening of trade and the intensification of competition between firms worldwide have spawned in recent economic behavior that were soon in conflict, at least in the short term, with the goals and the political problems of governments. In particular, three major centers have gradually found themselves in direct competition for influence and markets: Japan and, by extension, Southeast Asia, Europe, and North America. The debates on the triad as economic confrontation field have given to the agenda the question of the role of the state. Curiously, when the state is attacked from all sides, some forces pushing for a vision of the state strategist (White, 1993).

Even in the United States, the US government's actions are clearly in the service of US companies or those that behave in "American citizens." The confrontation between Europe, the United States, and China reveals, firstly, the importance of the issues and, secondly, the difficulties involved in reconciling without involving diplomacy and, therefore, the traditional relations between nations. The well-being in the short- and long-term economic is perceived to go through economic domination of some over others. Diamond concept we have seen, suggests that success depends on building a series of factors and conditions, in a given industry. It should be noted that the situation in any industry depends on companies, customers or suppliers, which are upstream and downstream of it. It is useful to think in groups based on industries related to each other. The idea of industry cluster and complements the old name "die" time used by economists, especially in France.

A cluster is a set of industries that are related. Implicitly, it is assumed that the health of one depends on the health of others. That's when governments began to take an interest not only to an industrial sector, as part of the formal diamond, but a set of sectors, trying to encourage the development of the most promising clusters for countries (Won and Lefevre, 1993). It suggests that a country must strengthen or abandon whole areas of the economy if it wants to become attractive to the field of dynamic companies. Ohmae, who was interested in the competition between the countries of the triad, suggested that the new form of strengthening of competitiveness is one that is explored or exploited by the Chinese government with developments by city or region.

In times of globalization, the two main external forces that shape business strategy are: the dynamics of the industry and the bargaining power of states. A sound global strategy can ignore neither the one nor the other. A sound strategy should also take into account the capabilities and enterprise resources, as the source of competitive advantage. The dynamics of a global industry is described in great detail by theories that treat oligopolistic competitive advantages and those dealing with foreign investment (Rugman, 1990). We resume here that what serves our purpose, which is to show how the dynamics of the industry strength recognizable strategic behavior. The dynamics of competition in an industry is generally determined by the following main factors (Doz, 1986): economies of scale; economies of experience; localization economies; the bases of the differentiation; nature of technology; distribution channels and export; access to capital.

You should also know that there are several types of economies of scale, the most important being those related to production, distribution, and customer service. Economies of scale in production are interacting with technological developments. They can be facilitated but also challenged by them. In general, there is enough stability in the development of manufacturing equipment for economies of scale, even if they are ultimately likely to be reconsidered, can first be decisive and favor companies that benefit the first.

Take, for example, the petrochemical industry. In the ranges of Capa, most common cities, when building production plants, a 2X capacity plant would cost only 20–50% more than a plant capacity X. Thus, in 1990, a single oil refinery, with a capacity of 90,000 barrels per day, costing 400 million dollars, while the same type of plant with a capacity twice as large, did not cost more than 600 million. These considerable economies of scale exist in all major manufacturing industries such as automotive, pulp and paper, steel, etc. It must, however, mention that most scales, the greater rigidities and inertia of the system are large, which can cause significant management costs and transaction that cancel the benefits of scale. This is why the emergence of small but solid competitors in traditionally dominated by economies of scale industries. The scale effect can also be cancelled or substantially affected by opportunities for differentiation.

Economies of scale, when the problems of management and sales are controlled, force decisions that can transcend the boundaries of a single country. Thus, in the automotive industry (Abernathy and Ginsburg, 1980), the convergence of tastes and needs among the countries of the triad (the United States, Europe, Japan), as well as the rationalization and compatibility between models introduced by businesses, has made the agenda the idea of "efficient size of production units." For example, General Motors had developed a standard size for the production of modules that contribute to the manufacture of a vehicle, which could go from 300,000 units to assemble a million for some molded equipment. As few national markets can support such large capacities, it is then necessary that these modules were built for several countries at once. Moreover, it is clear that the optimal use of these capabilities requires coordination between different national subsidiaries for product development, engineering, product introduction dates, and plant extensions.

Specialization is then inevitable, but it can be done if the states do not disrupt the coordination required between the specialized modules. It is therefore, in practice, a compromise between the requirements of States in the employment and research and development on the one hand, and the requirements of specialized production on a large scale, on the other. Thus, economies of scale in production are forcing plant specialization and multinational integration. Economies of scale in distribution of matter can also force specific behaviors. Or take the example of the automotive industry. Consider that for a given brand, a market share of 4–5% is required for a distribution network and dense enough service to be maintained. Consequently, economies of scale in the distribution plan will allow for wide or limited coverage strategies (concentrated) market. This is likely to be economies of distribution of matter in scale that have thwarted Renault France's efforts to become firmly established in North America.

In the case of the auto industry, economies of scale in production and distribution can have contradictory effects: the distribution economies of scale create pressures for the proliferation of models and their rapid replacement (to maintain consumer interest); the manufacturing economies of scale encourage rather high volumes for each model over several years. R & D costs also push manufacturers to large volumes per model. The experience effect can be described technically as the constant percentage decrease in costs for each doubling of production. This percentage is generally known for each type of industry or product and usually varies between 15 and 25%.

These savings are due to the increased expertise of employees (at all levels), which comes from the renewed production of the same product. Some of the savings are due to the gradual improvement of design and design, applicable to all facilities. However, much of these savings are related to location (where the business is located). In other words, they would not be available if a new plant was built elsewhere. This effect is therefore encouraging specialization and strengthening the scale effect. Also, differences in the cost factors are sources of benefits and savings for companies who can benefit.

In particular, this is true when one is able to locate intensive activities within a factor in countries that have an advantage in this factor. So, the countries of South Asia and Southeast were able, thanks to a low-wage workforce quality, attract production in many industrial sectors. These countries have also set up various mechanisms putting attract multinationals, the free trading zone 35 one can mention. It is clear, however, that if all countries offer the same benefits, localization economies may be lower. Similarly, the technology, including automation, tends to reduce the importance of the advantage of the "cost of labor" in some locations. These countries have also set up various mechanisms putting attract multinationals.

The distinction may seem the antithesis of globalization. In fact, it is common internationally. It may be based on the particular type of customer, the type of product and local tastes. For example, in the field of skiing, the market is homogeneous across the globe; differentiation is therefore made according to the skill and the size of the skier, giving a series of global segments that are often operated by different competitors. Differentiation can also be the result of clever marketing, including for some consumer products, especially when the approaches can be transferred from one market to another. This happens for example in the grain market, where Kellogg manage to distinguish in the minds of consumers, and in the beer,

Differentiation may also be based on very strong local and regional characteristics, which reduces its applicability globally. Thus, food in general, local tastes are usually very specific and force a distinction between different markets. Yet, some companies manage to stand out globally through massive advertising, but also to local adaptations; in the case of Nestlé Nescafe. Differentiation when promotes local content, opposes multinational integrations, particularly in terms of production, although it is not necessarily inconsistent with global harmonization of marketing expenses, research and development, and super general structure. International segmentation, however, has considerable advantages, and firms devote much of their creativity to recognize and exploit the opportunities segmentation applicable to several countries. Every day, new or clever advertising convergences allow to succeed where all seemed very local. For example, nobody would have believed that Kellogg's or McDonald's to take root so easily in France and worldwide, from Montreal to Casablanca via Moscow and the Champs Elysees.

Generally, in intensive industries working in the field of high technology (the ratio of sales allocated to spending on research and development is high) and companies that can spread the costs over larger volumes of production have a certain advantage. It is also clear that the internal technology is more likely to produce an advantage over those who do not have access, which shows the importance of spending on research and development. The technology, interacting with economies of scale, can promote greater specialization and integration activities. First, the volumes allow specialization can encourage the adoption of more advanced and efficient technologies. On the other hand, technological changes have often increased the economic size of the facilities. Doz (1986) mentions how the introduction of nuclear technology in the electricity industry has upset not only the production industry, but also production equipment significantly increasing plant size of 600 MW over 1000 MW.

Distribution channels are becoming important in at least four cases: (1) it must have its own channels; (2) when the flexibility of supplies is important; (3) when the channels are dominated by a small group of companies; and (4) when the tasks of service and sales are intense. Export channels are often expensive. The most powerful companies such as multinationals are more able to meet these costs and develop their own channels, while smaller domestic companies are forced to use agents or importers. However, domestic companies can sometimes more easily take advantage of government support. The nature of the sale may also encourage some strategies. Where distribution channels are easy to penetrate, they can promote a strategy of domination that connects several national markets; by cons, if the channels are controlled by manufacturers, most local strategies are favored. When there are no permanent channels, as in the sale of aircraft or sale of electric generation plants to developing countries, domestic companies can be successful in competing against multinationals.

When the sale involves intensive interaction with a fragmented customer or if it requires intensive services, promotes local businesses more familiar with the community, and is able to meet its needs. This raises the issue of access to capital: a presence in many markets can facilitate access to capital and can even reduce the cost of capital. Indeed, investors often have preference for foreign currency debt to diversify their portfolios. In conclusion, these dynamics appear to favor three types of strategies: (1) strategies that emphasize linkages between countries (including production linkages); (2) strategies that exploit the distinctiveness of each national market; and (3) strategies that attempt to take advantage of the homogenization of tastes and needs around the world.

In this regard, the competitive advantage of a company in a globalizing world context comes from its ability to provide resources compatible with the dynamics of the industry. In particular, a global company faces the need to respond to either local dynamics or to global and general dynamics for an entire industry or to global dynamics but on a specific need within a sector industrial. The company responds by using functional resources (general administration, production, marketing, finance, research and development, etc.) which are concentrated in a single country or who are dispersed in several countries. However, it is useful to clarify these functional resources by linking them to the value chain.

The concentration or dispersion of the elements of the value chain provides many interesting combinations. First, not only the elements can be made in different regions or countries, with production in a country, research and development in another, etc., but each element of the value chain can also be dispersed or concentrated. So, Toyota has long concentrated its production and most of its functions in Japan, dispersant that the marketing function to ensure the sale of its products in international markets. It is also, as is the case for companies such as Corning Glass Works (Corning called since 1989), a specialist technology glass product, distribute most of the activities of the value chain,

When the degree of globalization of the business is very large, each activity of the value chain is generally dispersed. Thus, if we take IBM, the activities of the chain directly involved in the creation of value are divided, but all support activities, including administrative infrastructure activities are also dispersed. For example, research and development are entrusted to a series of centers located everywhere in the world, including France, Germany, Italy, the United Kingdom, Japan, India, Canada and, of course, in the United States. Also, each region makes its own financial management, leaving the center for the international fund management movements. In general, the center is mainly responsible for international coordination. So, the possibilities of concentration or dispersion used to respond uniquely to the dynamics of competition. For example, an industry in which economies of scale are large, especially in production, will tend to favor a concentration of production activities or dispersion of these activities with a strong central coordination. Similarly, in an industry where the possibilities of global differentiation are large, the concentration of production activities may be required. Furthermore, if the nature of the product requires intensive customer relations, so it is necessary to disperse the marketing activities. An industry in which economies of scale are large, especially in production, will tend to favor a concentration of production activities or dispersion of these activities with a strong central coordination. Similarly, in an industry where the possibilities of global differentiation are large, the concentration of production activities may be required. Furthermore, if the nature of the product requires intensive customer relations, so it is necessary to disperse the marketing activities.

It is worth mentioning the terminology proposed by Prahalad and Doz (1987) because it largely coincides with that we have adopted so far. They use two dimensions: the degree of need for integration and the degree of need for local sensitivity (responsiveness). These two dimensions are then used, as we also do, to appoint, to clarify and assess the various strategic situations that may arise. Strategic definition proposed here and in the sections which follow, is only the first step of development of competitiveness in an industry globalization. A critical resource in the management of resources is the general management capacity, including the ability to direct and coordinate activities worldwide.

The matrix of Prahalad and Doz, linking integration and local sensitivity, suggests some basic strategies. These strategies are benchmarks rather than real strategies. In practice, the strategies will be a unique combination of integration and local sensitivity. However, this tells us that, even if the combination possibilities are numerous, even infinite in number, there are patterns that were often held in the literature. As for us, we will use the two dimensions of "dynamic industry" and "dispersion of the value chain" to establish some common combinations that allow consistency between the industry dynamics and characteristics of the value chain. When the dynamics of the industry is global, it often means that the economies of scale and experience are important. Similarly, the location of savings and the possibilities of access to capital can be numerous and easy to use by major players such as multinationals.

A global dynamic is often a stimulus to the technological developments that facilitate and sometimes even boost. Of course, global integration is possible only if the stakeholders have minimal control over distribution channels.

If, moreover, the value chain can be easily dispersed because the company has the resources and, among others, the managerial capacity to do so, then we are in a situation where the conditions favor a strategy where the system is facing a maximum cost reduction and therefore toward global integration. The typical example is that of large Japanese and American car companies, but also large computer companies like IBM or Microsoft. In the same industry globalization conditions, but when the value chain can be dispersed either because the company has neither the resources nor sufficient managerial capacity, either because the technology is sensitive and should be concentrated and protected, then the only possible strategy becomes an export strategy. This is often the case for companies still in the process of globalization or who have not yet gained sufficient confidence internationally.

As examples can be mentioned the textile and clothing companies such as Daewoo was in its creation (Aguilar, 1990), or as is the costume Peerless Clothing (Bonneau, 1995). The aircraft construction industry is another case. Markets are global, but the technology is sensitive and binding. Competition in different segments is global, and the shareholders are powerful; yet essentially, exports are dominant, even if, for ease, leading enterprises to agree to terms that promote local subcontracting. When the globalization of industry conditions are not yet together, especially with economies of scale, experience and location that does not disadvantage domestic players only.

The national sensitivity is also encouraged by the pressures of governments, and therefore the relationships that develop between competing firms, and incentives or penalties that governments put in place. In this case, when the resources and the company's capacity is large enough to allow the dispersion, often part of the value chain, one would think that the national sensitivity is the most appropriate strategy. Location decisions can be negotiated according to the benefits that the company can withdraw from the relationship with local governments. When the dynamics are not yet global and the value chain cannot be dispersed, (although sometimes it can be replicated in different places), one can speak of a multi-domestic strategy and a concentration strategy. The multidomestic strategy is when the company is reproduced locally without the possibility of linking activities at the international level; it is surprising to some technologically and geopolitically sensitive products, such as telecommunications, hazardous to health or the products taxed as cigarettes.

We speak of concentration strategy when capacity and corporate resources do not allow global coverage, as Entre State made working in the field of tobacco or alcohol.

The challenge of managing a global company is directly related to the strategy it gives. It is clear that the difficulties in managing the concentration or dispersion of the value chain are at the heart of the competitiveness of the company. These difficulties were already fairly discussed in the chapter on managing a complex organization. We consider here that issues related to the complexity engendered by the collapse in the world.

We can say that the international complexity poses three main types of problems: (1) a traditional, permanent problem effective implementation of the strategy, which generally ensures the convergence of efforts of the various subsidiaries; (2) an occasional problem of change in the relations between subsidiaries and senior management to meet the changing direction needs; (3) a flexibility problem keeping order to take advantage of opportunities that may arise or to meet unexpected difficulties in the implementation of the chosen direction.

These problems do not arise in the same way in all organizations. The strategic choices, as was discussed earlier, largely determine the nature of management required for implementation. In general, however, the management of a complex international organization is always a kind of meta-management. This meta-management is conducted using tools that are similar, regardless of the chosen strategy. It is the combination of these tools will change according to the strategy. To analyze these tools, it is useful to group according to the major challenges facing global companies, which generates three categories of tools: (1) tools for managing information and data systems that influences management behavior; (2) the tools to solve confrontations and conflicts that arise from the need to bring together shareholders who have different perspectives and who are exposed to different realities; (3) tools to directly manage the behavior of managers.

As we mentioned, the effect of the use of these tools can be felt in the short or long term. So everything related compensation and the situation of managers, including their powers and responsibilities, has a short-term effect. These tools are usually more concrete. Anything that affects the future situation of managers and their integration with the philosophy, way of behavior and traditions of the organization will have long-term effects. These tools have a cognitive and symbolic score. Moreover, it should be clear that the tools and actions of management information and data are technical and do not require constant involvement of senior management. On the other hand, direct management of managers and conflict resolution require attention and commitment of every moment. This is also the main task of the high direction organizations with significant international operations. As we said, the use of these tools is not the same as we need to integrate the system or being sensitive to local realities and pressures. Without addressing all situations, we refer to these extremes, the most important challenges and the means generally used to cope. The use of these tools is not the same as we need to integrate the system or being sensitive to local realities and pressures.

Multinational integration requires paying close attention to cost reduction and thus optimization of production and logistics system, while not neglecting the attention to be customer needs and market changes. It also raises significant problems in terms of relationships between headquarters and subsidiaries and on the relations between the local authorities and the company. An integrated production means among other things that factory output is closely interconnected. Thus, in the case of automotive companies, all components must be produced in time for the assembly to take place. Furthermore, production must be harmonized with market requirements. This requires considerable sophistication of the system and a complex international logistics management and high precision. However, there must be room for flexibility, as the system is open to exogenous shocks, including from the market.

Questions expansion or contraction of the system is even more complex. We must first assess the effects on the whole system, which is already not easy, but we must also reconcile the needs of efficiency and effectiveness with the requirements of the most important governments for business. Finally, it must be remembered that international investments still account for more risk and uncertainty due to exchange rate issues and changes in the relative position of factors from one country to another. In the case of an integrated company, the nature of the research and development can also make the problem management. Globalization of production can, under pressure from the government, force the globalization of research and development, which is not necessarily favorable to technological development. We meet so often in situations where the research and development system should require coordination as important as the production-logistics system with all the problems this can cause.

Thus, IBM maintenance of many research and development centers worldwide, facilitate integration. The company has experienced often, particularly by entrusting "missions" (design and starting innovation) and "controls" (development marketing) permanent to its most important centers. Thus, outside the United States, the four major centers of research and development in Germany, France, Japan, and the United Kingdom each have a special responsibility. For IBM, or automotive companies or chemistry, in integration voltages are then inevitable. Indeed, too much sensitivity to local realities can question the need for effective integration, while too much integration can affect market penetration.

Integration is seen with great suspicion by national governments, who see it as an attack on the nation's decision-making autonomy. They are subscribed until they understand the need of the industry dynamics and when they have developed confident and relaxed relationship with the companies in question. This is why the diplomatic efforts that integrated multinationals tend to do. For example, Ford has created, in the 1980s, a European advisory board, composed of individuals influencing EU policy. The other problem concerns the relationship with the unions, which have deteriorated because of the burst of integrated business systems. Trade unions are often in conflict with each other to support or oppose the decisions of production and investment of these companies. Governments are then often forced to intervene diplomatically to resolve disputes, but they do it differently. The Spanish government, for example, a large direct influence, the French government rather plays the role of referee, while the government of the United Kingdom intervenes very little.

The biggest problem remains the ability to measure and assess the performance of the subsidiaries. This is disrupted when interdependencies are many and involve the transparency of contributions. The accounting system must help to increase the visibility of local actions and consequences to prevent local prejudices disrupt the overall strategy. Doz (1986, p. 180) described some solutions implemented by large multinationals. The problems with the management of national sensitivity are generally well known. One can mention these: the need for all the risks and allocate resources among the subsidiaries; the need to avoid duplication in R & D, and spread the costs of it on a larger volume of activities; export coordination subsidiaries that produce the same equipment or products; technology transfer and, in general, know-how in all areas.

Resource allocation is disturbed largely because of different accounting standards, different investments, and taxation regulations differ from one country inflation rate to another, forcing many companies to accept coordination less than perfect. Corning Glass Works has many imperfections generally accepted in the enterprise systems where national sensitivity was required. Its interventions to ensure better coordination (Case Corning Glass Works, 1987) have generally not been successful. It remains that the leaders of multinationals in situations of national sensitivity are trying to ensure consistency in behavior. In particular, they use the following instruments:

- Some consistency of information systems, especially for planning, budgeting, and control. Note that the allocation of resources remains an act of confidence in the abilities and judgment of subsidiary leaders, underscoring the importance of personalities in the management of such a strategy.

- Encouraging corporate behavior through measurement and systems of rewards and punishments. For this, we tend to measure performance based on the objectives set by the subsidiaries themselves rather than standard measuring instruments, less necessary here only when there is integration.
- The peer review, which can be done when peers are involved both in the design, with collective responsibilities, and evaluation of plans and requests for funds and other resources. This is done by creating "business teams or products" for the coordination and responsibility of the profits.
- The direct involvement of senior management. This may be precisely in operations, as did Geneen (1984), or more vaguely on behavior, as did the president of Schlumberger Riboud.

After this overview, we can imagine what to do so that there are no problems in the management of a large global company. It is probably necessary to demonstrate a great genius in strategic design and development of the required management instruments and not to forget that even though everyone tries, few succeed. This probably means that the tools and analysis that we discussed so far are not sufficient. These are the ingredients of a truly artistic nature of construction. This construction assumes that leaders are able of leadership and there are also mechanisms to all go in the right direction. These mechanisms are of symbolic, ideological, and cultural.

Conclusion

Throughout this book, we are interested in understanding meander of the applied management. Traditionally, the management of the organization is exercised by the executive or by a team of leaders. Following external and internal diagnostics, the leader establishes consistency between the elements it needs, wants or can consider. Once strategy is formulated, it is communicated to other levels of the organization, and implementation tools, primarily a structure and appropriate management processes, facilitate its implementation.

This process is often presented in a linear way from formulation to implementation. It is also presented as one-time, localized in time, and lying before the action. Despite some problems, among others, the unpredictability of the future and limited cognitive abilities of leaders, this process is possible in smaller organizations operating in relatively stable industries. But when the size of the company increases, the organization becomes complex, as in the case of the diversified businesses, or the context becomes more complex, as in the case of global companies, the officer often need be helped to properly undertake its maker business.

It is in this context very well described by Ansoff (1965), what is operational planning. It also is in this context that Porter systematic approaches such as the industrial economy have become increasingly important in decision optimization.

All of these tools to establish internal and external diagnostics business are useful to the leader and his management team. This is why the traditional approach to business strategists is still widespread, both in organizations and in business schools. This also explains the importance we have given to the planning of activities throughout the book.

This part of the book focuses on how leaders can conduct an orderly operational thinking, and discusses several tools that can help them realize the analysis of the business context and the organization, and make strategic choices appropriate. Having ordered activity of strategic thinking, analyze and make operational choices consistent with the elements of the analysis, plan activities and programs based on these choices: this systematic procedure has many advantages. It leads the company to discipline one to think of the future; it forces him to have goals to reach and organize action to achieve these objectives; it provides the basis for evaluating the performance of organizational units and individuals; it creates order in the organization and is safe for executives, managers, and employees.

But there are also problems with this approach, problems that have been experienced by several companies that have been identified by several authors in the late 1960s (Loasby, 1967; Steiner, 1972; Wildavski, 1973; Morgan, 1983), which were summarized by Mintzberg (1994). For Mintzberg, a major problem lies in the fact that this process can disempower the leader leaving too much room to analysts. However, there are other problems that seem more important.

First, an approach to training strategies which implies that the company's leaders fail of the strategic competence of other members of the organization and the contribution they can make. On the other hand, an approach that wants a formalized process taking place mainly before the action, may not be sensitive to strategies that may emerge during the action. It is then necessary to design the exercise of the strategist art can take another form. It is no longer for the leader to first formulate a strategy and then implement it. It is rather to facilitate the strategic action of all members of the organization and to establish an environment that enables their contribution to training strategies.

It is more a one-off activity, but an action that is done in small steps, "along the way" in the words of Avenier (1997). And it is this process of this strategic action daily strategy emerges. Such an approach does not eliminate the role of leaders, on the contrary. As we have seen, the leaders remain the architects of the rationale and context of creators, and these roles are both important and demanding. It did not evacuate the importance of intentional strategies. These can be an element of the strategic action around what action of players develops; they can no longer be regarded as THE business strategy. This strategic action leads to daily strategy.

This approach to training strategies is particularly relevant in situations of complexity, turbulence, and instability, when it is difficult to understand the operational context in which the organization operates, and the skills to cope. Some approaches attempt to reduce the complexity of the context of business. This is what Porter chooses to focus primarily competitive economic environment and reduce strategic diversity in three generic strategies. Other approaches recognize the complexity, but put forward simple rules that companies must adopt if they want to deal with it. This is what Eisenhardt and Sull explain in their article "Strategy as Simple Rules" (2001).

In both approaches we have just mentioned, it was the same with the tools we have presented on the strategic management of diversified business and global business. It is still mainly the leaders who are responsible to find solutions that allow the company to operate and be successful in complex situations.

In addressing the training strategies through strategic action of all members of the organization, we offer a different path. The strategy is a social construct, involving all members of the organization. They have a strategic competence flowing of learning they have acquired and tacit and explicit knowledge they have acquired over time. This knowledge is then put to use on a daily basis to solve problems that occur or for projects of all kinds.

This is the daily work of all shareholders that form the organization's strategy. Avenier (1997) prefers to talk about groping strategy. This strategy differs from incremental strategy, since it may allow radical changes. It is also different from the emerging strategy because it allows the realization of "deliberate actions in emergent situations"; it also promotes "the emergence of deliberation, that is to say the emergence of deliberate projects." The tentative strategy is therefore built step by step, through multiple oscillations between reflection and action in a permanent tension between deliberate and emergent. It allows the organization to adapt and to adjust continuously in a changing business context: by multiple oscillations between reflection and action in a permanent tension between deliberate and emergent. As we mentioned earlier, the leaders continue to play important roles, but they no longer act as the only designers of the strategy. As architects of the reason, they ensure that the shared representations correspond to the values and major objectives of the organization; they also ensure that the procedures and routines are consistent with this system of representations. As creators of context, they set up a context for participation and innovation, and they ensure that the structure and different management systems make possible participation and creativity. It is through their leadership that leaders create an environment that enables members of the organization to participate actively and creatively, the strategic action of the company.

Thanks to their leadership that all members of the organization have to work together in the same direction, so that the company performs economically and socially. Designed in this way, the strategy formation process becomes a sense of co-construction process, shared by the different levels of the organization, and a co-management process "along the way." The analysis and action become interrelated processes; the action can no longer be conceived as a simple implementation of strategic choices made at the top of the organization; so the company performs economically and socially.

References

Abell, D.; Hammond, J. S. *Strategic Market Planning*; Prentice-Hall: Englewood Cliffs, NJ, 1979.

Abernathy, W.; Ginsburg, D. *Government, Technology and the Future of the Automobile*; McGraw-Hill: New York, 1980.

Abernathy, W.; Wayne, K. Limits of the Learning Curve. *Harvard Bus. Rev.* **1974**.

Aguilar, F. J. *General Managers in Action*; Oxford University Press: New York, 1988.

Allaire, Y.; Firsirotu, M. How to Implement Radical Strategies in Large Organizations. *Sloan Manage. Rev.* **1985,** *26.*

Allison, G. *The Essence of Decision Explaining the Cuban Missile Crisis*; Little Brown: Boston, MA, 1971.

Amedzro St-Hilaire, W. Context of Using Reliability and Validity Measures: Relevance of Correlation in the Study of Causation. *Manage. Appl. Econ. Rev.* **2009,** *15* (1).

Amedzro St-Hilaire, W. Regulation and Corporate Governance: Role of Deficiencies of Boards in the Bankruptcy of Industrial and Commercial Crown Corporations. *Res. Sci. Gest: Manage. Sci.: Cien. Gest.* **2010,** *80.*

Amedzro St-Hilaire, W. *L'adaptation Organisationnelle dans les Théories Managériales et Sociales;* Ed. Presses de l'Université du Québec, 2011a, 167 pp.

Amedzro St-Hilaire, W. *Management des Télécoms d'États en Mutation: Défis et Démarche Opérationnelles;* Ed. Presses de l'Université du Québec, 2011b, 187 pp.

Amedzro St-Hilaire, W. *Global Environment of Contemporary Public Action;* Ed., North American Business Press, 2011c.

Amedzro St-Hilaire, W. A Systemic Approach of the Strategic Score in the Modern Organization System Model. *Acta Univ. OEcon.* **2012,** *8* (5).

Amedzro St-Hilaire, W. Le Renouvellement des Techniques Économiques de Gestion et la Maîtrise des Finances dans le cas des Firmes Multinationales. *Rapport d'étude Financé par Multilateral Investment Guarantee Agency.* World Bank Group, 2013.

Amedzro St-Hilaire, W. Operational Risk Control & Project Effectiveness In Strategic Project Management. *Rev. Gest. 2000* **2014,** *31* (5).

Amedzro St-Hilaire, W. Méthodes et Techniques d'Analyses de Marchés Industriels en Phase de Préinvestissement. *Rapport d'étude financé par Multilateral Investment Guarantee Agency,* 2015a.

Amedzro St-Hilaire, W. *Perspective Stratégique et Gestion Opérationnelle de l'Économie Bancaire;* La Coll. Perspective Organisationnelle, L'Harmattan, 2015b, 261 pp.

Amedzro St-Hilaire, W. Structural Breaks in the Interconnected Relation between Decision-Making and Applied Strategic Agility on Public Insurance and Bank Projects. *Strat. Change: Brief. Entrepreneur. Finance* **2016a,** *25* (4).

Amedzro St-Hilaire, W. Risk Planning, Takeover and Stabilizing Performance in Applied Strategic. *Strat. Change: Brief. Entrepreneur. Finance* **2016b,** *25* (3).

Amedzro St-Hilaire, W. *Sciences de Gestion et Pratiques Managériales: La Rationalisation de la Planification Stratégique Face à la Dynamique Concurrentielle des Institutions Bancaires et Financières.* Working Paper. Université Paris 1 Pantheon Sorbonne, 2016c.

Amedzro St-Hilaire, W. *La Gestion Économique de Projets Industriels et Technologiques;* Ed. L'Harmattan, 2016d, 218 pp.

Amedzro St-Hilaire, W. *La Gestion Stratégique des Projets Énergétiques;* Ed. L'Harmattan, 2016e, 276 pp.

Amedzro St-Hilaire, W. *La Dynamique Institutionnelle;* Coll. Logiques Sociales, L'Harmattan, 2017a, 219 pp.

Amedzro St-Hilaire, W. *La Gestion Planifiée: Logique, Fondement et Perspectives;* Ed. L'Harmattan, Col. Dynamiques d'Entreprises, 2017b, 260 pp.

Amedzro St-Hilaire, W. Optimizing the Empirical Foundations of the Temporal Organization for Managerial Model Efficiency: A Multiple Readiness Framework of Determinants? *Rev. Quest. Manage.* **2018,** *21.*

Amedzro St-Hilaire, W. The Strategic Impact of Performance Appraisal on Corporate Governance Dynamics. *J. Manage. Dev.* **2019a,** *38* (5).

Amedzro St-Hilaire, W. *Entrepreneurship Strategies & Policies;* Coll. Business Management; Ed. Apple Academic Press, CRC Press, and Taylor & Francis Group, 2019b, 314 pp.

Amedzro St-Hilaire, W. *La Formation Endogène de Coalitions dans Les Jeux Avec Externalités et Les Démarche Opérationnelles Communes Des Lobbyistes Rassemblés au Sein de "réseaux" Pour Influencer l'Attribution des Aides au Développement et les Politiques de Marchés Publics.* Working Paper. Université Paris 1, Pantheon Sorbonne, 2019c.

Amedzro St-Hilaire, W.; Boisselier, P. Evaluating Profitability Strategies and the Determinants of the Risk Performance of Sectoral and Banking Institutions. *J. Econ. Admin. Sci.* **2018,** *34* (3).

Amedzro St-Hilaire, W.; Boisselier, P. The Coordinated Strategy for the Optimization of the Interaction Level of Business Model. *J. Econ. Admin. Sci.* **2019,** *35* (1).

Amedzro St-Hilaire, W.; Chiasson, G. *Une Approche Sémantique des Démarche Opérationnelles Opérationnelles de la Gestion Institutionnelle et Sectorielle.* Working Paper. Centre de Recherche sur la Gouvernance des Ressources Naturelles, Université du Québec en Outaouais, 2014.

Amedzro St-Hilaire, W.; De La Robertie, C. The Influence of IT Innovation Mechanisms on Healthcare Project Management Strategies. *J. Econ. Bus.* **2017,** *31* (1).

Amedzro St-Hilaire, W.; De La Robertie, C. Correlates of Affective Comment in Organizational Performance: Multi-Level Perspectives. *Austr. J. Career Dev.* **2018a,** *27* (1).

Amedzro St-Hilaire, W.; De La Robertie, C. The Empirical Impacts of Risk-Management on the Enhancement of Control Mechanisms in Organizational Studies. *Res. Sci. Gest.: Manage. Sci. Cien. Gest.* **2018b,** *123.*

Amedzro St-Hilaire, W.; Prasad, B. The Governance Correlation Matrix of the Financial Institutions. *Manage. Appl. Econ. Rev.* **2018,** *26* (1).

Anderson, D. Une Démarche pour Revitaliser les Grandes Entreprises. *Rev. Franç. Gest.* **1986.**

Andrews, K. *The Concept of Corporate Strategy;* Richard-Irwin: Homewood, IL, 1971.

Andrews, K. *The Concept of Corporate Strategy*; Richard-Irwin: Homewood, IL, 1987.

Ansoff, H. *Corporate Strategy: An Analytic Approach to Business Policy for Growth and Expansion*; McGraw-Hill: New York, 1965.

Arrègle, J. Analyse Resource Based et Identification des Actifs Opérationnelle. *Rev. Franç. Gest.* **1996,** *Mars/Avril/Mai*, 25–36.

Astley, W.; Fombrun, C. Collective Strategy: The Social Ecology of Organizational Environment. *Acad. Manage. Rev.* **1983,** *8*, 576–586.

Avenier, M. *La Démarche Opérationnelle Chemin Faisant*; Economica: Paris, 1997.

Baker, R. Goodbye to Newspapers? *NY Rev Books* **2007,** *54* (13).

Balogun, J.; Johnson, G. Organizational Restructuring and Middle Manager Sense-Making. *Acad. Manage. J.* **2004,** *47* (4), 523–549.

Barnard, C. *The Functions of the Executive*; Harvard University Press: Cambridge, MA, 1968.

Bartlett, C., Ghoshal, S.; Doz, Y. Global Strategic Management Impact on the New Frontiers of Strategy Research. *Strat. Manage. J.* **1991,** *12*, numéro spécial, été.

Bartlett, C., Ghoshal, S.; Doz, Y. What is a Global Manager?. *Harvard Bus. Rev.* **1992,** *Sept.–Oct.*

Baughman, J. *Problems and Performance of the Role of Chief Executive of the General Electric Company, 1892–1974*; texte ronéotypé, Harvard Business School, 1974.

Bhambri, A.; Greiner, L. New CEO Intervention and Dynamics of Deliberate Strategic Change. *Strat. Manage. J.* **1989,** *10*, 67–86.

Biggadike, E. *Corporate Diversification: Entry, Strategy and Performance*; Harvard University Press: Cambridge, MA, 1979.

Blanc, M. Pour un État Stratège Garant de l'intérêt Général. *Rapport Commission Parlementaire sur le Rôle de l'État*; Documentation Française: Paris, 1993.

Borys, B.; Jemison, D. Hybrid Arrangements as Strategic Alliances Theoretical Issues in Organizational Combinations. *Acad. Manage. Rev.* **1989,** *14*, 234–249.

Bower, J. *Managing the Resource Allocation Process*; Irwin: Homewood, IL, 1970.

Bower, J. *The Two Faces of Management: An American Approach to Leadership in Business and Politics*; Houghton Mifflin: Boston, MA, 1983.

Braybrooke, D.; Lindblom, C. A Strategy of Decision: Policy Evaluation as a Social Process; Collier Macmillan: Londres, 1979.

Bresser, R.; Harl, J. Collective Strategy: Vice or Virtue. *Acad. Manage. Rev.* **1986,** *11*, 408–427.

Bresser, R. Matching Collective and Competitive Strategies. *Strat. Manage. J.* **1988,** *9*, 375–385.

Burgelman, R. A Process Model of Internal Corporate Venturing in the Diversified Major Firm. *Admin. Sci. Quarterly* **1983,** *28* (2), 223–244.

Burgelman, R. Intraorganizational Ecology of Strategy Making and Organizational Adaptation: Theory and Field Research. *Organ. Sci.* **1991,** *2* (3), 24–56.

Caves, R.; Jones, R. *World Trade and Payments: An Introduction*; De Boeck University Press: Bruxelles, 1981.

Chakravarthy, B. Measuring Strategic Performance. *Strat. Manage. J.* **1986,** *7*, 437–458.

Chandler, A. *Strategy and Structure*; MIT Press: Cambridge, MA, 1962.

Chandler, A. *The Visible Hand*; Harvard University Press: Cambridge, MA, 1977.

Chandler, A. The Evolution of Modern Global Competition. In *Competition in Global Industries*; Porter, M. E., Ed.; Harvard Business School Press: Boston, MA; 1986; pp 405–448.

Chandler, A. *Scale and Scope*; Harvard University Press: Cambridge, MA, 1990.

Chenery, H.; Srinivasan, T. *Handbook of Development Economics;* Elsevier: Amsterdam, 1998; pp 1442–1480.

Child, J. Organization, Structure, Environment and Performance: The Role of Strategic Choice. *Sociology* **1972,** *6* (1), 1–22.

Christensen, C. R., Andrews, K. R.; Bower, J. L. *Business Policy, Texte et Cas*; Irwin: Homewood, IL, 1973.

Christiansen, L. Les Démarche Opérationnelles de Diversification dans les Vagues de Fusions et d'Acquisitions des Années 1980. *Rev. Int. Gest.* **1987,** *12* (3).

Collins, J.; Porras, J. *Built to Last*; Harper Business: New York, 1994.

Côté, M. *La Gestion Stratégique d'Entreprise*; Concept et cas: Gaëtan Morin éditeur, 1991.

Côté, M. *La Gestion Stratégique: Aspects Théoriques*; Gaëtan Morin, 1995.

Crozier, M.; Friedberg, E. *L'acteur et le Système. Les Contraintes de l'Action Collective*; Ed. Seuil, 1977.

Cvar, M. Patterns of Globalization. In *Competition in Global Industries*; Porter, M. E., Ed. Free Press: New York, 1986.

Cyert, R.; March, J. A Behavioral Theory of the Firm; Prentice-Hall: Englewood Cliffs, NJ, 1963.

Dahl, R. The Concept of Power. *Behav. Sci.* **1957,** *2*, 201–215.

Doz, Y. The Evolution of Cooperation in Strategic Alliances: Initial Conditions or Learning Processes? *Strat. Manage. J.* **1986,** *17*, 55–83.

Drucker, P. *The Practice of Management*; Harper and Row: New York. Version française: (1957). La pratique de la direction des entreprises. Éditions d'Organisation: Paris, 1952.

Dunning, J.; Pierce, R. *Profitability and Performance of the World's Largest Industrial Companies*; The Financial Times: Londres, 1985.

Dutton, J.; Jackson, S. Categorizing Strategic Issues Links to Organizational Action. *Acad. Manage. Rev.* **1987,** *12* (1).

Eisenhardt, K.; Sull, D. Strategy as Simple Rules. *Harvard Bus. Rev.* **2001,** *79* (1), 106–116.

Evan, W. The Organization-Set: Toward a Theory of Interorganizational Relations; Maurer, dans J. G.; In *Readings in Organization Theory: Open-System Approaches*; Random House: New York, 1966.

Fligstein, N. The Intraorganizational Power Struggle Rise of Finance Personnel to Top Leadership in Large Corporations, 1919–1979. *Am. Sociol. Rev.* **1987,** *52*, 44–58.

Frederickson, J.; Iaquinto, A. Inertia and Creeping Rationality in Strategic Decision Processes. *Acad. Manage. J.* **1989,** *32*, 516–542.

Fréry, F. *Les Réseaux d'Entreprises: Une Approche Transactionnelle*; Laroche, dans H.; Nioche, J.-P., Ed.; Repenser la démarche opérationnelle, Paris: éditions Vibert (Entreprendre, série Vital Roux), 1998.

Fry, J.; Killing, J. *Canadian Business Policy: A Casebook*; Prentice-Hall: Scarborough, ON, 1983.

Gagliardi, P. The Creation and Change of Organizational Cultures: A Conceptual Framework. *Organ. Stud.* **1986,** *7* (2), 117–134.

Gagné, P.; Lefèvre, M. *L'Entreprise à Valeur Ajoutée: Le Modèle Québécois*; Publi-Relais: Paris, 1993.

Galbraith, J. Strategy and Organization Planning. *Hum. Resour. Manage.* 1983, printemps-été.

Galbraith, J.; Nathanson, D. *Strategy Implementation: The Role of Structure and Process*; West Pub: Saint Paul, MN, 1978.

Geertz, C. *The Interpretation of Cultures*; Basic Books: New York, 1973.

Geneen, H. *Managing, Garden City*; Doubleday: New York, 1984.

Ghoshal, S.; Bartlett, C. Linking Organizational Context and Managerial Action: The Dimensions of Quality of Management. *Strat. Manage. Rev.* 1994, *15*, 91–112.

Gioia, D.; Chittipeddi, K. Sensemaking and Sensegiving in Strategic Change Initiation. *Strat. Manage. J.* 1991, *12* (6), 433–448.

Gomez, P. *Le gouvernement de l'Entreprise*; Inter Éditions: Paris, 1996.

Greenwood, R.; Hinings, C. Organizational Design Types, Tracks and the Dynamics of Strategic Change. *Organ. Stud.* 1988, *9*, 293–316.

Hambrick, D.; Mason, P. Upper Echelons: The Organization as a Reflection of Its Top Managers. *Acad. Manage. Rev.* 1984, *9*, 193–206.

Hamel, G.; Doz, Y.; Prahalad, C. *Collaborate with Your Competitors and Win*; University Press: Cambridge, MA, 1989.

Hampden-Turner, C. La Culture d'Entreprise: Des Cercles Vicieux Aux Cercles Vertueux; Seuil: Paris, 1992.

Hannan, M.; Freeman, J. *Organizational Ecology*; Harvard University Press: Cambridge, MA, 1989.

Harrison, D.; Laplante, N. Confiance, Coopération et Partenariat. Un Processus de Transformation dans l'Entreprise Québécois. *Relat. Ind.* 1994, *49* (4), 696–729.

Heckscher, E. The Effect of Foreign Trade on the Distribution of Income. *Ekon. Tidskr.* 1919, 497–512.

Hedberg, B. How Organizations Learn and Unlearn. In *Handbook of Organizational Design*; Nystrom, C., Starbock, W. H., New York: Oxford University Press, 1982; vol 1, pp 3–27.

Helleiner, G. K. *Transnational Corporations and Direct Foreign Investment*, 1989.

Henderson, L. J. *On the Social System: Selected Writings*; University of Chicago Press: Chicago, IL, 1970.

Hofstede, G. *Culture's Consequence International Differences in Work-Related Values*; Sage: Beverly Hills, 1980.

Inkpen, A.; Choudhoury, N, The Seeking of Strategy Where It Is Not: Towards a Theory of Strategy Absence. *Strat. Manage. J.* 1995, *16*, 313–323.

Jauch, L.; Glueck, W. *Strategic Management and Business Policy*; McGraw-Hill: New York, 1988.

Johnson, G. Rethinking Incrementalism. *Strat. Manage. J.* 1988, *9*, 75–81.

Katz, D.; Kahn, R. *Social Psychology of Organizations*; Wiley: New York, 1966.

Kets de Vries, M.; Miller, D. *Organization on the Couch: Clinical Perspectives on Organizational Behavior and Change*; Jossey Bass: San Francisco, CA, 1991.

Kim, D. The Link between Individual and Organizational Learning. *Sloan Manage. Rev.* 1993, 37–50.

Laroche, H.; Nioche, J. L'Approche Cognitive de la Démarche Opérationnelle. *Rev. Franç. Gest.* **1994,** *Juin–Juil–Août.*

Lawrence, P.; Lorsch, J. *Organizations and Environment*; Irwin: Homewood, IL, 1967.

Levitt, T. Marketing Myopia. *Harvard Bus. Rev.* **1960,** *Juil.–Août,* 45–56.

Lindblom, C. The Science of Muddling Through. *Publ. Admin. Rev.* **1959,** *19,* printemps.

Loasby, B. Long-Range Formal Planning in Perspectives. *J. Manage. Stud.* **1967,** *IV,* 300–308.

Lovas, B.; Ghoshall, S. Strategy as Guided Evolution. *Strat. Manage. J.* **2000,** *21* (9), 875–896.

Malkiel, B. *Random Walk Down Wall Street: The Time-Tested Strategy for Successful Investing*; W. W. Norton & Company: New York, 2003.

March, J. G.; Simon, H. A. *Organizations*; Wiley: New York, 1958.

March, J.; Olsen, J. *Ambiguity and Choice in Organizations*, Universitetsforlaget: Bergen, 1976.

March, J. Bounded Rationality. *Bell J. Econ.* **1978,** *9* (2).

March, J.; Olsen, J. *Rediscovering Institutions: The Organizational Basis of Politics*; Free Press: New York, 1989.

McGahan, A. *Note on Industry and Company Profitability*; Note 9-793-139, Harvard Business School: Boston, MA, 1993, 1999.

Miller, D. *The Icarus Paradox*; Harper Business: New York. Version française: 1992. Presses de l'Université Laval: Le paradoxe d'Icare, Sainte-Foy, QC, 1990.

Miller, D.; Floricel, S. *Games of Innovation in the New and Old Economy: An Exploratory Study of the Networks of Value Creation and Capture.* Article présenté à l'Academy of Management Conference, Hawaii, 2005.

Miller, D.; Shamsie, J. The Resource-Based View of the Firm in Two Environments: The Hollywood Film Studios from 1936 to 1965. *Acad. Manage. J.* **1996,** *39* (4).

Mintzberg, H. Patterns in Strategy Formation. *Manage. Sci.* **1978,** *24,* 934–948.

Mintzberg, H. Crafting Strategy. *Harvard Bus. Rev.* **1987,** *Août–Sept.*

Mintzberg, H. Strategy Formation—Schools of Thought. In *Perspectives on Strategic Management*; Fredrickson, dans J. W., Ed.; Harper & Row: New York, 1990; pp 105–235.

Mintzberg, H. *Grandeur et Décadence de la Planification Stratégique*; Dunod: Paris, 1994.

Mintzberg, H.; McHugh, A. Strategy Formation in an Adhocracy, *Admin. Sci. Quarterly* **1985,** *30.*

Mintzberg, H.; Waters, J. Of Strategies: Deliberate and Emergent. *Strat. Manage. J.* **1985,** *6,* 257–272.

Morgan, G. Rethinking Corporate Strategy: A Cybernetic Perspective. *Hum. Relat.* **1983,** *36* (4), 345–360.

Morris, D.; Hergert, M. Trends in International Collaborative Agreements. *Columbia J. World Bus.* **1987,** 15–21.

Nantel, J. La Segmentation, un Concept Analytique Plutôt Que Stratégique. *Gestion* **1989,** *14* (3).

Noël, A. Les Modèles de Décision en Acquisitions et Fusions. *Gestion* **1987,** numéro spécial, Septembre.

Noël, A. Strategic Cores and Magnificent Obsessions Discovering Strategy Formation through Daily Activities of CEOs. *Strat. Manage. J.* **1989,** *10,* 33–49.

Nonaka, I. A Dynamic Theory of Organizational Knowledge Creation. *Organ. Sci.* **1994,** *5* (1), 14–37.

Ohlin, B. *Interregional and International Trade*; Harvard University Press: Cambridge, MA, 1933.

Ohmae, K. *Triad Power the Coming Shape of Global Competition*; Free Press: New York, 1985.

Palich, L., Cardinal, L.; Miller, C. Curvilinearity in the Diversification: Performance Linkage: An Examination of Over Three Decades of Research. *Strat. Manage. J.* **2000,** *21* (2).

Parsons, T. *Structure and Process in Modern Societies*; Free Press: New York, 1960.

Pascale, R. Perspectives on Strategy: The Real Story behind Honda's Success. *Calif. Manage. Rev.* **1984,** *26*, 47–72.

Pascale, R.; Athos, A. Le Management est-il un Art Japonais? Éditions d'organisation: Paris, 1984.

Pascale, R. *Managing on the Edge*; Touchstone: New York, 1990.

Peters, T.; Waterman, R. *Le Prix de l'Excellence*; Inter Éditions: Paris, 1983.

Pfeffer, J.; Salancik, G. The External Control of Organizations: A Resource Dependence Perspective. Harper and Row: New York, 1978.

Porac, J. On the Concept of Organizational Community. In *Evolutionary Dynamics of Organizations*; Baum J., Singh, J., Ed.; Oxford University Press: New York, 1994; pp 451–456.

Porter, M. *Industry Structural Change*. Note 9-377-051; Harvard Business School: Boston, MA, 1976.

Porter, M. *Competitive Strategy Techniques for Analyzing Industries and Competitors*; Free Press: Toronto, 1980.

Porter, M. *Choix Opérationnelle et Concurrence*; Economica: Paris, 1982.

Porter, M. E. *Competitive Advantage*; Free Press: New York, 1985.

Porter, M. E. *L'Avantage Concurrentiel*; Inter Éditions: Paris, 1986.

Porter, M. *Competiting in Global Industries*; Harvard Business School Press: Boston, 1987.

Porter, M. *The Competitive Advantage of Nations*; Macmillan: Londres, 1990.

Porter, M. Competing Strategy Revisited: A View from the 1990s: *The Relevance of a Decade*, Harvard Business School Press, 1994.

Prahalad, C.; Bettis, R. A. The Dominant Logic: A New Linkage between Diversity and Performance. *Strat. Manage. J.* **1986,** *7*, 485–501.

Prahalad, C.; Doz, Y. The Multinational Mission Balancing Local Demands and Global Vision. Free Press: New York, 1987.

Prahalad, C.; Hamel, G. Strategic Intent. *Harvard Bus. Rev.* **1989,** *Mai–Juin*, 63–76.

Prahalad, C.; Hamel, G. The Core Competence of Corporation. *Harvard Bus. Rev.* **1990,** *Mai–Juin*, 79–91.

Prahalad, C.; Hamel, G. *Competing for the Future*; Harvard Business School Press: Cambridge, MA, Version française (1995), La conquête du futur, Inter Éditions: Paris, 1994.

Quinn, J. *The Intelligent Enterprise, a Knowledge and Service-Based Paradigm*; Free Press: New York, 1992.

Reger, R., Gustafson, L.; Mullane, J. Reframing the Organization: Why Implementing Total Quality Is Easier Said the Done. *Acad. Manage. Rev.* **1994,** *19* (3), 565–584.

Rieger, F.; Wong-Rieger, D. Model Building in Organizational/Cross-Cultural Research: The Need for Multiple Methods, Indices, and Cultures. *Int. Stud. Manage. Organ.* **1988,** *18*, 19–32.

Roethlisberger, F. *The Elusive Phenomena*; Harvard University Press: Cambridge, MA, 1977.

Rowe, A., Mason, R.; Dickel, K. *Strategic Management and Business Policy: A Methodological Approach*; Addison Wesley: Reading, MA, 1982.

Rugman, A. *Research in Global Strategic Management*; JAI: Greenwich, CT, 1990; vol 1.

Rumelt, R. How much Does Industry Matter? *Strat. Manage. J.* **1991,** 12 (3).

Salter, M.; Weinhold, W. *Diversification through Acquisition*; Free Press: New York, 1979.

Schein, E. Organizational Culture and Leadership, Jossey-Bass: San Francisco, 1985.

Schein, E. What Is Culture? In *Reframing Organizational Culture*; Frost, P.; Moore, L., Louis, M., Lundberg, C., Martin, J., Eds.; Sage: Newbury Park, CA, 1991; pp 243–253.

Schein, E. Culture: The Missing Concept in Organization Studies; *Admin. Sci. Quarterly* **1996,** *41* (2), 229–240.

Schendel, D.; Hofer, C. Strategic Management A New View of Business Policy and Planning, Boston: Little Brown, 1978.

Schwenk, C. Cognitive Simplification Processes in Strategic Decision-Making. *Strat. Manage. J.* **1984,** *5*, 111–118.

Scott, W. *Institutions and Organizations*; Sage: Thousand Oaks, CA, 2001.

Selznick, P. *Leadership in Administration*; University of California Press: Los Angeles, CA, 1957.

Sikora, M. Merger Disclose Rules Are Eased. *Merg Acquis* **2000,** *35*, 12.

Simon, H. *Administrative Behavior. A Study of Decision-Making Processes in Administrative Organization*; Macmillan: New York, 1945, 1947, 1957, 1973, 1975, 1976.

Sloan, P. *Placer Dome's Policy on Sustainability: Program Business and the Environment*; Schulich School of Business—York University: Toronto, 1999.

Smircich, L.; Stubbart, C. Strategic Management in an Enacted World. *Acad. Manage. Rev.* **1985,** *10*, 724–736.

Spender, J. *Industry Recipes: An Enquiry into the Nature and Sources of Managerial Judgement*; Basil Blackwell: Cambridge, 1989.

Steiner, C. *Pitfalls in Comprehensive Long-Range Planning*; The Planning Executives Institute: Oxford, 1972.

Stigler, G. Capitalism and Monopolistic Competition: I. The Theory of Oligopoly. *Am. Econ. Rev., Pap. Proc.* **1950,** *40*, mai.

Stevenson, H. Defining Corporate Strengths and Weaknesses. *Sloan Manage. Rev.* **1976,** printemps.

Stonich, P. Using Rewards in Implementing Strategy. *Strat. Manage. J.* **1981,** *2*, 345–352.

Stratégor. Démarche Opérationnelle, Structure, Décision, Identité. Inter Éditions: Paris, 1988.

Thiétart, R.-A. La Démarche Opérationnelle d'Entreprise. In *Organizations in Action*; Thompson, J. D., Ed.; McGraw-Hill: New York, 1990.

Thorelli, H. Networks: Between Markets and Hierarchies. *Strat. Manage. J.* **1986,** *7*, 37–51.

Tichy, N. *Managing Strategic Change: Technical, Political and Cultural Dynamics*; John Wiley & Sons: New York, 1983.

Tunstall, W. Breakup of the Bell System: A Case Study of Cultural Transformation. In *Gaining Control of the Corporate Culture*; Jossey-Bass: San Francisco, 1985; pp 44–65.

Tushman, M.; Romanelli, E. Organizational Evolution: A Metamorphosis Model of Convergence and Reorientation. *Res. Organ. Behav.* **1985,** 7, 171–222.

Vandangeon-Demurez, I. La Dynamique des Processus de Changement. *Rev. Franç. Gest.* **1998,** *Sept./Oct.*, 120–138.

Vernon, R.; Wells, L. *Economic Environment of International Business*; Prentice-Hall: Englewood Cliffs, NJ, 1976.

Von Bertalanffy, L. *General System Theory*; Braziller: New York, 1968.

Weick, K. *The Social Psychology of Organizing*; Addison-Wesley: Reading, MA, 1979.

Weick, K.; Westley, F. Organizational Learning: Affirming an Oxymoron. In *Handbook of Organization Studies*; Clegg, dans S., Hardy, C., Nord, W., Ed.; Londres: Sage, 1996; pp 440–458.

Wernefelt, B. A Resource Based View of the Firm. *Strat. Manage. J.* **1984,** 5, 171–180.

Westley, F. Middle Managers and Strategy: Microdynamics of Inclusion. *Strat. Manage. J.* **1990,** 11, 337–351.

Westley, F.; Mintzberg, H. *Profiles of Strategic Vision: Levesque and Iacocca. Charismatic Leadership*; Jossey-Bass: San Francisco, CA, 1988.

Wildadski, A. If Planning Is Everything, Maybe It's Nothing. *Policy Sci.* **1973,** 4, 127–153.

Woodward, J. *Industrial organization: Theory and Practice*; Oxford University Press: Londres, 1965.

Index